ER FOR ALGEBRA

ER FOR ALGEBRA

KEVIN TUBBS

To order additional copies of this book, contact:
Xlibris Corporation
1-888-795-4274
www.Xlibris.com
Orders@Xlibris.com
64733

CONTENTS

11-16   Useful terms

17-28   Simplifying Numerical Expressions by Using the Order of Operations

29-30   Evaluating Algebraic Expressions Using Substitution and the Order of Operations

31-33   Simplifying Algebraic Expressions by Using the Order of Operations

34-49   Solving Linear Equations Containing One Variable

50-58   Using Formulas

59-66   Solving Linear Inequalities Containing One Variable

67-71   Solving Linear Inequalities That Require Multiplying or Dividing Both Sides of the Inequality by a Negative

72-87   Graphing Relations

88-103   Linear Equations in Two Variables

104-117   Systems of Linear Equations in Two Variables

118-126   Solving Quadratic Equations Containing One Variable

127-128   Translating Verbal to Math

129-141   Word Problems

142   Quick Review of Adding Integers

143   Quick Review of Subtracting Integers

144   Quick Review of Fractions

145-146   Quick Review of Powers

147   Solving Proportions

Introduction

This manual is designed to help individuals who have been exposed to elementary algebra and **want some assistance** with understanding and execution. For the examples to be effective, **reading carefully** is required. There is **no easy route or shortcut** to understanding algebra. Sorry.

The text will demonstrate processes used to solve problems. The processes are shown step by step and explanations are given for why each step is made. Make sure to understand one step before proceeding to the next. Make sure to gain understanding of the changes that are made from one step to another. For example, if one line contains the expression 2+3 and the next line doesn't, make certain to determine where the expression went. It was probably replaced with 5 since 2+3 = 5. Reading a math text requires time and the **willingness to focus** and to **reread** statements to gain understanding.

Work hard and enjoy the experience of learning!

Kevin

Information Needed

Throughout the text there are tables that show the order in which problems are to be completed. The tables must be read from the top left box to the right. After completing a row, go to the left box in the next row and again follow the boxes to the right. Continue the pattern until the problem is complete.

Step 1	Step 2	Step 3	Step 4
Step 5	Step 6	Step 7	Step 8
Step 9	Step 10	Step 11	Step 12

The best strategy is to gain understanding of one step prior to reading the next. However, there will be times when reading ahead is beneficial.

Useful terms

The following are descriptions that will help with understanding math terms. Do not consider the following descriptions dictionary worthy definitions.

Algebraic expression—Mathematical phrase that can contain numbers, variables and algebraic operation symbols.

Algebraic operations—Addition, subtraction, multiplication, division and roots such as square roots.

Area—A numerical measurement that expresses the two dimensional size of a plane or curved surface is the area.

>   **Surface area**—A measurement used to represent the sum of the areas of the outer parts of a three dimensional object.

Coefficient—The numerical factor in a term. The coefficient of $3x^2$ is 3.

Coordinate Plane—A plane that is structured with a vertical number line (y-axis) and a horizontal number line (x-axis) which intersect at 0. The plane is used to graph relations (sets of ordered pairs). The number lines are numbered by ones unless otherwise noted.

Coordinate Plane

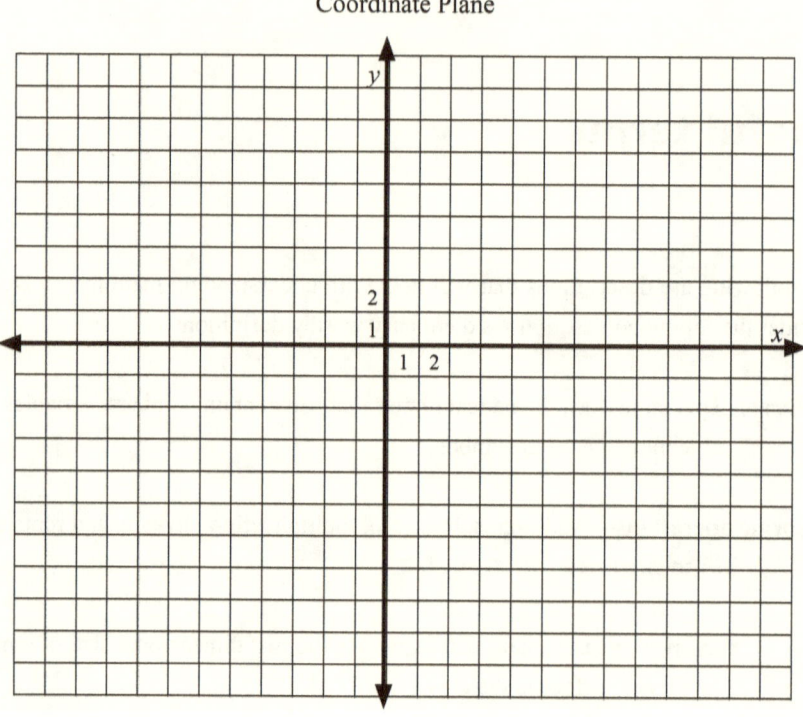

Dependent variable—The second coordinate in an ordered pair. The output of a relation is dependent.

Distributive Property—$a(b + c) = ab + ac$. The factor a can be multiplied by the two terms within the factor $b + c$ to eliminate parenthesis in the expression.

Division—Multiply by the reciprocal. $a \div b = a \cdot \dfrac{1}{b}$

Examples: $5 \div \dfrac{1}{2} = 5 \cdot \dfrac{2}{1} = 5 \cdot 2 = 10$ $\qquad \dfrac{2}{3} \div \dfrac{3}{5} = \dfrac{2}{3} \cdot \dfrac{5}{3} = \dfrac{10}{9}$

Domain—The set of all first coordinates of a relation.

Equation—A mathematical sentence containing an equal (=) sign. The sentence says that algebraic expressions represent the same number.

Linear Equation—In one variable, say x, an equation that can be written as $ax + b = 0$. a and b represent constants and a can not equal zero. In two variables, say x and y, $ax + by + c = 0$. a, b and c are constants. a and b can not both be zero. Continue with three variables etc.

Quadratic Equation—In one variable, say x, an equation that can be written in the form $ax^2 + bx + c = 0$. a, b and c are constants and a can not equal zero. There must be a variable raised to the second power. Quadratic equations can have multiple variables as well.

Formula—Symbols used to represent mathematical relationships.

Example: $A = bh$ (area equals base times height)

Quadratic Formula—Used to solve quadratic equations in one variable. To solve for x in $ax^2 + bx + c = 0$, use the formula: $x = \dfrac{-b \pm \sqrt{b^2 - 4ac}}{2a}$

Function—A set of ordered pairs, or relation, where every first coordinate is paired with exactly one second coordinate. Functions are often represented with rules that establish a relationship between the two coordinates. (See relation.) The first coordinate is often called the input value, while the second coordinate is called the output value.

Examples: 1. {(2,3), (4,5), (7,9), (-1,3), (-2,5), (8,9)} No first coordinates repeated.

2. Rule $y = 2x - 4$ 3. Rule $f(x) = 2x^2 + 4x + 5$

Independent variable—The first coordinate in an ordered pair. The input value in a function is independent.

Inequality—Linear in one variable, say x, is a mathematical sentence that provides comparisons of algebraic expressions, usually providing order.

Examples: $2x < 5$ $3x - 1 \geq 2$ $4 > 5x + 3$

Inverse Operations—Operations that "undo" each other are inverse operations. The inverse of addition is subtraction and vice versa. The inverse of multiplication is division and vice versa.

Opposites—Numbers are opposites if their sum is zero. The opposite of 5 is -5 because $5 + (-5) = 0$.

Order of Operations—The standard order to perform operations when simplifying or evaluating expressions used to assure consistent values.

First do work within grouping symbols.
Next do powers.
Next do multiplication and division from left to right.
Then, do addition and subtraction from left to right.

Perimeter—The length of the boundary of a closed two dimensional shape is the perimeter.

Point-slope form—Given a point on a line (x_1, y_1) and the slope of the line m, the equation $y - y_1 = m(x - x_1)$ is the point-slope form of the equation of the line.

Power—The notation, a^n, used to represent the product of a used as a factor n times.

Example: 5^3 means $5 \cdot 5 \cdot 5$.

Base (of a power)—In the power a^n, the number a is called the base.

Exponent (of a power)—In the power a^n, the number n is called the exponent.

Proportion—An equation that states **ratios** are equal.

$$\frac{a}{b} = \frac{c}{d}$$

Range—The set of all the second coordinates of a relation.

Rate—A ratio that compares two values with different units.

Ratio—An expression that compares numbers by division is a ratio. $\frac{a}{b}$

Reciprocals—Reciprocals are two numbers whose product is 1. 3 and $\frac{1}{3}$ are reciprocals because $3 \cdot \frac{1}{3}$ is 1. $-2\frac{2}{5}$ and $-\frac{5}{12}$ are reciprocals since their product is 1.

Relation—A relation is simply a set of ordered pairs. Relations can be represented by equations (rules), graphs, tables and maps.

Examples: 1. {(2,3), (4,5), (7,9), (2,3), (-2,5), (8,9)}   2. Rule y = 2x -4

3. Table   4. Map

x	y
2	6
-3	2
2	1
5	8

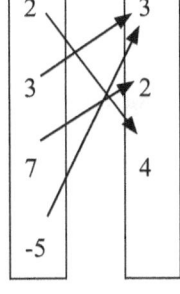

5. Graph

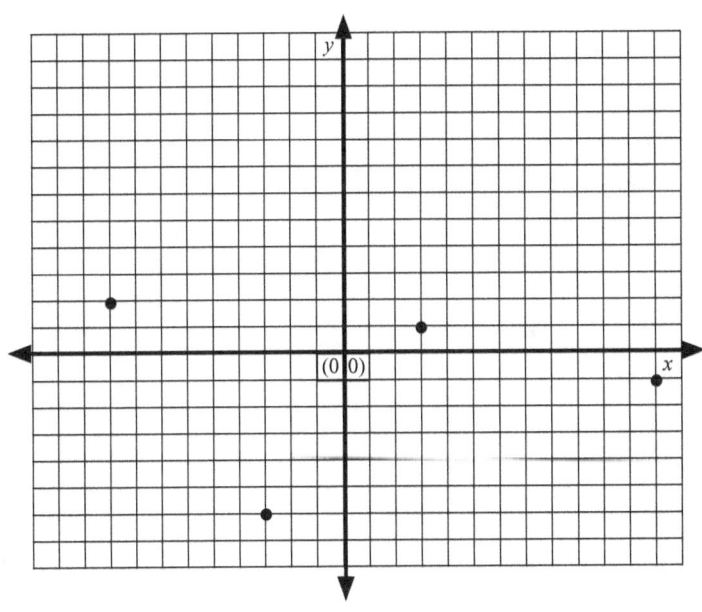

Domain (of a relation)—The set of all the first coordinates of a relation.

Range (of a relation)—The set of all the second coordinates of a relation.

Slope (rate of change)—The slope of a line, m, with points (x_1, y_1) and (x_2, y_2) is given by the formula below. The slope is a number that is used to describe the steepness of a line.

$$m = \frac{y_2 - y_1}{x_2 - x_1}$$

Example:

The slope of the line that contains the points (2, 5) and (-4, 7) is found by the following process.

Use one point as (x_1, y_1) and the other as (x_2, y_2). Then, substitute into the formula

$$m = \frac{y_2 - y_1}{x_2 - x_1}.$$

$$m = \frac{7 - 5}{-4 - 2} = \frac{2}{-6} = -\frac{1}{3}$$

Slope-intercept form—The slope-intercept form for the equation of a line is y = mx + b where m is the slope and b is the y-intercept of the line.

Standard form—The standard form for the equation of a line is Ax + By = C Where A, B and C are constants, with A and B, not both zero.

Subtraction—Subtraction means "add the opposite." a - b = a + (-b)

Examples: 11 - 5 = 11 + (-5)   -12 - 14 = -12 + (-14)
-13 - (-2) = -13 + 2

Variable—A symbol, normally a letter, used to represent a set of numbers or a number.

x-**intercept**—The *x*-intercept in a plane is the *x*-coordinate where the graph of a relation, intercepts the *x*-axis. Find the *x*-intercept by replacing *y* with zero and solving for *x*.

y-**intercept**—The *y*-intercept in a plane is the *y*-coordinate where the graph of a relation intercepts the *y*-axis. Find the *y*-intercept of a relation by replacing *x* with zero and solving for *y*.

Simplifying Numerical Expressions by Using the Order of Operations

An algebraic expression is a phrase that consists of numbers, variables and algebraic operations. Below are examples of algebraic expressions.

$$2 + 2(6 - 4) - 21$$
$$w + 3$$
$$3r - 2r + 17$$
$$11 \cdot 15 - (3 + 6)$$
$$2n - 7 + 3 (n - 4)$$
$$-\frac{10}{5} - 11$$
$$2^3 + 2 (4 - 6)^2$$

Notice that some of the expressions have variables (letters or symbols used to represent numbers) and some don't. When an algebraic expression has no variables, it is called a numerical expression. Simplifying in this case means finding the number the expression represents or equals. For example, 2+2 represents the number 4. When the expression 2+2 is simplified, the value is 4.

Algebraic expressions with variables can be evaluated if the numbers the variables represent are given. For example, $2r + 3$ can be evaluated if the value of r is given. Say the value of the r is 3. In other words $r = 3$. Then, $2r + 3$ is replaced with $2(3) + 3$. The value of the expression is then 9, since $2(3) = 6$ and $6 + 3 = 9$.

There is an accepted order to follow when simplifying, or evaluating expressions. The order is called the **Order of Operations**. Following this order is **required** whenever expressions are being simplified or evaluated! The **Order of Operations** tells which operation should be executed first, second, third and so on until the expression is simplified.

The **Order of Operations** is below.

First do work within grouping symbols.
Next do powers.
Next do multiplication and division from left to right.
Then, do addition and subtraction from left to right.

Here are examples of using the **Order of Operations**.

Grouping symbols. Grouping symbols include parenthesis ( ), brackets [ ], braces { }, radicals $\sqrt{}$ and the fraction bar $\overline{}$.

The expression $11^2 - 3 - 4(5 - 11)$ contains a power, subtraction, subtraction, multiplication and subtraction in grouping symbols. The order of operations requires that the subtraction in parenthesis, 5 - 11, be completed first.

$$11^2 - 3 - 4(5 - 11) =$$
$$11^2 - 3 - 4(-6)$$

Powers. Powers are expressions that involve exponents. The expression 3^4 is read three to the fourth power and means $3 \cdot 3 \cdot 3 \cdot 3$. The 3 is called a base and the 4 is called an exponent. The exponent tells how many times to use the base as a factor.

The expression $11^2 - 3 - 4(-6)$ contains a power, subtraction twice and multiplication. The power must be done first by the order of operations.

$$11^2 - 3 - 4(-6) =$$
$$121 - 3 - 4(-6)$$

Multiplication and division from left to right. Remember that the choice between doing multiplication or division first is determined by which operation is to the left.

For example, simplifying $2 \div 4 \cdot 3$ requires that 2 be divided by 4 prior to multiplying by 3. $2 \div 4(3) = 0.5(3) = 1.5$

The expression 121 - 3 - 4(-6) contains subtraction, subtraction and multiplication. The multiplication must be done first by the order of operations.

$$121 - 3 - 4(-6) =$$
$$121 - 3 - (-24) =$$
$$121 - 3 + 24$$

Addition and subtraction from left to right. Remember that the choice between doing addition or subtraction first is determined by which is to the left.

For example, simplifying 2 - 4 + 3 requires that 4 be subtracted from 2 prior to adding 3. $2 - 4 + 3 = -2 + 3 = 1$

The expression 121 - 3 + 24 contains subtraction and addition. The subtraction is done first by the order of operations. Addition and subtraction are done left to right.

$$121 - 3 + 24$$
$$118 + 24$$

Add the remaining values.   142

An example of using the **<u>Order of Operations</u>** is as follows.

Simplify $3(2 + 7)^2 - 5$. ➤ $3(2 + 7)^2 - 5$

The choices are multiplication by 3, parenthesis containing addition of 2 and 7, a power with exponent 2 and subtraction of 5. The addition is in parenthesis so, it is done first. The addition in parenthesis leads to

$3(9)^2 - 5$ (Where did the 9 come from?)

The choices left are multiplication, a power and subtraction. The power is done prior to multiplication. Executing the power leads to

$3 \cdot 81 - 5$ (Where did the 81 come from?)

The choices left are multiplication and subtraction. Multiplication is done next. After multiplying the expression is

$243 - 5$. (Where did the 243 come from?)

The last operation is subtraction. After subtracting the value of the expression is

238 (Where did the 238 come from?)

The next page shows how simplifying $3(2 + 7)^2 - 5$
should appear on paper without commentary.

Here is how simplifying $3(2 + 7)^2$ - 5 should appear on paper.

$$3(2 + 7)^2 - 5$$

$$3(2 + 7)^2 - 5$$
$$3(9)^2 - 5$$
$$3(81) - 5$$
$$243 - 5$$
$$238$$

The expression $3(2 + 7)^2$ - 5 equals the number 238.

Information Needed.

Throughout the text there are tables that show the order in which problems are to be solved. The tables must be read from the top left box to the right. After completing a row, go to the left box in the next row and again follow the boxes to the right. Continue the pattern until the problem is complete.

Step 1	Step 2	Step 3	Step 4
Step 5	Step 6	Step 7	Step 8
Step 9	Step 10	Step 11	Step 12

The best strategy is to gain understanding of one step prior to reading the next. However, there might be times when reading ahead is beneficial.

Order of Operations	Use the **Order of Operations**	The **Order of Operations** states	The choices left
<u>**Order of Operations**</u> 1. **First do work within grouping symbols.** 2. **Next do powers.** 3. **Next do division and multiplication from left to right.** 4. **Then, do addition and subtraction from left to right.**	Use the <u>**Order of Operations**</u> to simplify the expression below. 6 - 3 + 4 · 2 Always identify all the choices of operations to perform in the expression. The choices are subtraction, addition and multiplication.	The <u>**Order of Operations**</u> states that of the choices multiplication is done first. When the multiplication is done 6 - 3 + 4 · 2 is rewritten as 6 - 3 + 8. (Where did 8 come from?)	The choices left are subtraction and addition. 6 - 3 + 8 Since the subtraction is to the left of the addition, subtraction is done next. When the subtraction is completed, 6 - 3 + 8. is rewritten as 3 + 8.
The only operation left is addition. When the addition is completed, 3 + 8 is rewritten as 11.	So, 6 - 3 + 4 · 2 has value 11. 6 - 3 + 4 · 2 = 11	Here is the way the problem should appear on paper.	6 - 3 + 4 · 2 = 6 - 3 + 8 = 3 + 8 = 11 **Completed.**
Simplify the expression below. 6 + 8 ÷ 4 · 2 Remember to identify the choices of operations before starting.	**SHOW WORK**		**The solution is on the next page.**

Simplify

$6 + 8 \div 4 \cdot 2 =$ The choices are addition, division and multiplication. Division and multiplication are done prior to addition. The **Order of Operations** states that division and multiplication are done from left to right. Therefore, division is done first.

$6 + 2 \cdot 2 =$ The choices are now addition and multiplication. Multiplication is done next.

$6 + 4 =$ Finally, add.

10

The steps as they should appear on paper are shown below.

$$6 + 8 \div 4 \cdot 2 =$$
$$6 + 2 \cdot 2 =$$
$$6 + 4 =$$

$$10$$

Order of Operations	Use the **Order of Operations** to simplify the expression below.	(-12)(3) means to multiply -12 by 3. The parenthesis, in this situation, implies multiplication.	The choices left are multiplication, subtraction and multiplication.
1. **First do work within grouping symbols.**	$8(2 - 4) - (-12)(3)$		
2. **Next do powers.**	Always identify the choices of operations to perform.	The **Order of Operations** states that of the choices work in parenthesis must be done first.	The next step is a choice between the two multiplications. The multiplication on the left is done next.
3. **Next do division and multiplication from left to right.**	The choices are multiplication, $8(2 - 4)$ means 8 times the quantity $(2 - 4)$, parenthesis containing subtraction, subtraction and multiplication.	After subtracting inside the parenthesis $8(2 - 4) - (-12)(3)$ is rewritten as $8(-2) - (-12)(3).$	After multiplying $8(-2) - (-12)(3)$ is rewritten as $-16 - (-12)(3).$ (Where did -16 come from?)
4. **Then, do addition and subtraction from left to right.**			
The last choice is between subtraction and multiplication. Multiplication takes precedence. After multiplying, $-16 - (-12)(3)$ is rewritten as $-16 - (-36).$	Last, subtract and $-16 - (-36)$ is rewritten as 20. Remember $-16 - (-36)$ means $-16 + 36$ which equals 20.	$8(2 - 4) - (-12)(3)$ is equal to 20. $8(2 - 4) - (-12)(3) = 20$ Here is the way the problem should appear on paper.	$8(2 - 4) - (-12)(3) =$ $8(-2) - (-12)(3) =$ $-16 - (-12)(3) =$ $-16 - (-36) =$ 20 **Completed.**
Simplify the expression below. $3(6 - 4) - 2(-12 + 4)$ Remember to identify the choices of operations before starting.	**SHOW WORK**		**The solution is on the next page**

Simplify

3(6 - 4) - 2(-12 + 4) =   The choices are multiplication, parenthesis containing subtraction, subtraction, multiplication and parenthesis containing addition. The parenthesis to the left is done first. 6 - 4 =2

3(2) - 2(-12 + 4) =   The choices are now multiplication, subtraction, multiplication and parenthesis containing addition. The parenthesis is done next. -12 +4 = -8

3(2) - 2(-8) =   The choices are now multiplication, subtraction and multiplication. The multiplication to the left is next. 3(2) = 6

6 - 2(-8) =   The choices are now subtraction and multiplication. The multiplication is next. 2(-8) = -16

6 - (-16) =   Finally subtract. Remember that subtraction means add the opposite. 6 - (-16)= 6 + 6 = 22.

22

The steps as they should appear on paper are shown below.

3(6 - 4) - 2(-12 + 4) =

3(2) - 2(-12 + 4) =

3(2) - 2(-8) =

6 - 2(-8) =

6 - (-16) =

22

Completed.

Order of Operations 1. **First do work within grouping symbols.** 2. **Next do powers.** 3. **Next do division and multiplication from left to right.** 4. **Then, do addition and subtraction from left to right.**	Use the **Order of Operations** to simplify the expression below. $$\dfrac{16-5\cdot3+3(15)\div5}{6\div3\cdot4}$$ The fraction bar is a symbol that is used for grouping. When simplifying expressions, work the numerator and denominator separately. Then, use the fraction bar to divide.	$$\dfrac{16-5\cdot3+3(15)\div5}{6\div3\cdot4}$$ Always identify the choices of operations to perform. The choices in the numerator are subtraction, multiplication, addition, multiplication and division.	Multiplication and division are done prior to addition and subtraction. Since $5\cdot3$ is farthest left it is done first. $$\dfrac{16-5\cdot3+3(15)\div5}{6\div3\cdot4}=$$ $$\dfrac{16-15+3(15)\div5}{6\div3\cdot4}$$ Three times 15 is next.
$$\dfrac{16-15+3(15)\div5}{6\div3\cdot4}=$$ $$\dfrac{16-15+45\div5}{6\div3\cdot4}$$ (Where did 45 come from?) Next the division is executed. $$\dfrac{16-15+45\div5}{6\div3\cdot4}=$$ $$\dfrac{16-15+9}{6\div3\cdot4}=$$ $$\dfrac{1+9}{6\div3\cdot4}$$	$$\dfrac{16-15+9}{6\div3\cdot4}$$ (Where did 9 come from?) Between subtraction and addition, the operation to the left is done next. Subtract 15 from 16. $$\dfrac{16-15+9}{6\div3\cdot4}=$$ $$\dfrac{1+9}{6\div3\cdot4}$$	To complete the numerator, add $1+9$. $$\dfrac{1+9}{6\div3\cdot4}=$$ $$\dfrac{10}{6\div3\cdot4}$$ Now focus on the denominator.	$$\dfrac{10}{6\div3\cdot4}$$ Division, $6\div3$ is left of multiplication so it is done first. $$\dfrac{10}{6\div3\cdot4}=$$ $$\dfrac{10}{2\cdot4}$$ (Where did 2 come from?)
Next, multiply $2\cdot4$. $$\dfrac{10}{2\cdot4}=$$ $$\dfrac{10}{8}$$	$\dfrac{10}{8}$ can be placed in lower terms by dividing the numerator and denominator by 2. $$\dfrac{10\div2}{8\div2}=\dfrac{5}{4}$$	So, $$\dfrac{16-5\cdot3+3(15)\div5}{6\div3\cdot4}=\dfrac{5}{4}$$	**Following is what should appear on paper.**

Normally the steps taken to simplify the numerator and denominator are done at the same time. Keep this in mind while following the steps. The operations performed for each step are in the box to the right of the solution.	$$\frac{16-5\cdot3+3(15)\div5}{6\div3\cdot4}=$$ $$\frac{16-15+3(15)\div5}{2\cdot4}=$$ $$\frac{16-15+45\div5}{8}=$$ $$\frac{16-15+9}{8}=$$ $$\frac{1+9}{8}=$$ $$\frac{10}{8}=$$ $$\frac{5}{4}$$	Commentary: Numerator $5\cdot3$ Denominator $6\div3$ Numerator $3(15)$ Denominator $2\cdot4$ Numerator $45\div5$ Denominator **simplified** Numerator 16-15 Numerator $1+9$ Write the fraction in lowest terms by dividing numerator and denominator by 2. **Completed!**	
Simplify the expression below. $$\frac{4\cdot6-3(2-4)+12}{6-3+4}$$	**SHOW WORK**		**The solution is on the next page.**

Simplify.

$$\frac{4\cdot6-3(2-4)+12}{6-3+4}=$$

Numerator 2 - 4 is done first
Denominator 6 - 3 is done first

$$\frac{4\cdot6-3(-2)+12}{3+4}=$$

Numerator next multiply $4\cdot6$
Denominator next add 3+4

$$\frac{24-3(-2)+12}{7}=$$

Numerator multiply (3)(-2)
Denominator Simplified

$$\frac{24-(-6)+12}{7}=$$

Numerator subtract 24 -(-6) which equals
24 + 6
Denominator Simplified

$$\frac{30+12}{7}=$$

Numerator add 30+12
Denominator Simplified

$$\frac{42}{7}=$$

Divide to place in lowest terms

6

Completed

Cover the words and review the process
as it should be written without commentary.

Evaluating Algebraic Expressions Using Substitution and the Order of Operations

Algebraic expressions containing variables can be evaluated if the value of the variable is known. Consider the following examples.

Find the value of the expressions below with values of each variable given.
$$a = 3 \quad b = -7 \quad c = 4$$

Evaluate 3a.

Since the value of a is 3, replace a in the expression $3a$ with 3.
The expression then is $3 \cdot 3$ which is 9.
The value of $3a$, in this situation is 9. $3a = 9$

Evaluate 2a + b.

Since the value of a is 3 and the value of b is -7,
replace each variable with the number given.
The expression now is $-2 \cdot 3 + (-7)$
Use the order of operations to simplify the expression.
$-2 \cdot 3 + (-7) = -6 + (-7) = -13$
The value of $-2a + b$, in this situation is -13. $-2a + b = -13$

Evaluate $-b + ac$.

Notice the negative sign prior to b. The sign says to find the opposite of b.

When b is replaced with its value, be sure to write -(-7).

The expression after substituting is -(-7) + 3 · 4. Simplify.

$$-(-7) + 3 \cdot 4 = 7 + 12 = 19$$

The value of $-2a + b$ is 19 in this situation. $-2a + b = 19$

Order of Operations	Use the **Order of Operations**	$\dfrac{2x+6-y}{6 \div 3 \cdot x} =$	Always identify the choices of operations to perform.
1. **First do work within grouping symbols.**	to evaluate the expression below. $x = 5 \qquad y = -13$	$\dfrac{2 \cdot 5 + 6 - (-13)}{6 \div 3 \cdot 5}$	The choices in the numerator are multiplication, addition and subtraction.
2. **Next do powers.**	$\dfrac{2x+6-y}{6 \div 3 \cdot x}$	Pay particular attention to the substitution of -13 for y since y is negative.	Which operation should be done first?
3. **Next do division and multiplication from left to right.**	First replace the variables with their values.		
4. **Then, do addition and subtraction from left to right.**		Simplify.	
Following is the process of simplifying. Recall that the numerator and denominator can be simplified simultaneously. Make sure to evaluate each step.	$\dfrac{2 \cdot 5 + 6 - (-13)}{6 \div 3 \cdot 5} =$ $\dfrac{10 + 6 - (-13)}{2 \cdot 5} =$ $\dfrac{16 - (-13)}{10} =$	$\dfrac{29}{10}$ The value can also be expressed in the following ways. 2.9 or $2\dfrac{9}{10}$	

Simplifying Algebraic Expressions by Using the Order of Operations

In the prior examples the value of the expression was found. When expressions contain variables and the value of the variables are known, the expressions can be evaluated. There are algebraic expressions that can be simplified without being evaluated. If the expression contains variables with unknown values, then simplify without evaluating. For example, consider the expression below.

$$2w + 3w$$

The expression can be simplified by combining the two terms. Two terms that have the same variables and the same exponents on the variables are like terms.

$$2w + 3w =$$
$$5w$$

Another example of an expression that can be simplified without being evaluated is below.

$$3(4d - 5)$$

The expression is not simplified since the expression contains parenthesis. The **Distributive Property** is used to eliminate the parenthesis. This property tells that the factor 3, outside the parenthesis, can be multiplied by each term inside the parenthesis. 3 can be multiplied by 4d and 3 can be multiplied by 5. The process is as follows.

$$3(4d - 5) =$$
$$3 \cdot 4d - 3 \cdot 5 =$$
$$12d - 15 \quad \text{Simplified}$$

Simplified expressions contain no parenthesis and no like terms that are not combined.

The algebraic expression $2x^2 - 5 + 4x^2$ is not simplified. The expression contains like terms, $2x^2$ and $4x^2$. $2x^2$ and $4x^2$ are like terms because they have the same variables and the same exponents on the variables. The expression is simplified below.

$$2x^2 - 5 + 4x^2$$
$$2x^2 + 4x^2 - 5$$
$$6x^2 - 5 \quad \text{simplified}$$

Here is an example of simplifying an expression that can't be evaluated.

Simplify. $3(p + 7) - 4 + 2p$ The expression contains parenthesis. Therefore, it is not simplified. Clear the parenthesis by using the distributive property.	$3(p + 7) - 4 + 2p =$ $3 \cdot p + 3 \cdot 7 - 4 + 2p =$ $3p + 21 - 4 + 2p$ (Where did 3p come from?) The expression is not simplified because there are still like terms.	$3p$ and $2p$ are like terms. 21 and 4 are like terms. The terms can be rearranged; but be careful to keep the correct operation with the term. $3p + 21 - 4 + 2p =$ $3p + 2p + 21 - 4$	Next, combine like terms. $3p + 2p + 21 - 4 =$ $5p + 17$ (Where did 17 come from?) **The expression is simplified.** **Following is what should appear on paper.**
$3(p + 7) - 4 + 2p$ $3 \cdot p + 3 \cdot 7 - 4 + 2p$ $3p + 21 - 4 + 2p$ $3p + 2p - 4 + 21$ $5p + 17$	**Simplify.** $18 - 4(e - 7) - 4 + 5e$	**SHOW WORK**	**The solution is on the next page**
Simplify. $\dfrac{3}{4}(e - 4) + 4 - 5e$	**SHOW WORK**		**The solution is on the next page.**

Simplify.

$18 - 4(e - 7) - 4 + 5e$	Distribute fist.
$18 - 4 \cdot e - 4(-7) - 4 + 5e$	Multiply to simplify.
$18 - 4e + 28 - 4 + 5e$	Arrange to collect like terms.
$-4e + 5e + 18 + 28 - 4$	Combine like terms.
$e + 42$	Complete

Simplify.

$$\frac{3}{4}(e-4)+4-5e$$

$$\frac{3}{4}(e)-\frac{3}{4}(4)+4-5e$$

$$\frac{3}{4}(e)-5e-3+4 \quad \text{(see note below)}$$

$$-\frac{17}{4}(e)+1$$

Completed.

Note: To subtract $\frac{3}{4}(e) - 5e$, follow the steps. $\quad \frac{3}{4}(e)-5e=\frac{3}{4}(e)-\frac{20}{4}(e)=-\frac{17}{4}(e)$

Solving Linear Equations Containing One Variable

An equation is a statement that contains an equal sign stating that two expressions represent or equal the same number. Below are examples of equations.

$$2 + 2 = 4$$

$$w = -7 \qquad \text{linear}$$

$$3r = 17 \qquad \text{linear}$$

$$11 \cdot 15 = 165$$

$$2n - 7 = 21 \qquad \text{linear}$$

$$-\frac{t}{5} - 11 = 21 \qquad \text{linear}$$

$$k^2 + 3 = 12$$

Notice that some of the equations have variables in them. These are called open sentences and one of the numbers is represented by a variable. When there is only one variable that appears in the equation and the variable is to the first power (no exponent represents exponent 1), the equation is called a linear equation in one variable.

The goal in **solving** an equation is to find the number that the variable can be replaced with, that will make the equation a true statement. One process for solving equations involves "undoing" operations until the variable is isolated. For example, the statement $2x$ states that x is multiplied by 2. The expression will show only x when the $2x$ is divided by 2. Division will undo multiplication. $\frac{2x}{2} = x$.

Information Needed

Operations that undo each other are called inverse operations.

Division is the inverse of multiplication. For example, start with the number 2. Now multiply the 2 by 4 and the result is 8. Next divide 8 by 4 and 2, the original number, is the result. Dividing by 4 undoes multiplying by 4. Division undoes multiplication.

Multiplication is the inverse of division. For example, start with the number 12. Now divide the 12 by 4 and the result is 3. Next multiply 3 by 4 and 12, the original number, is the result. Multiplying by 4 undoes dividing by 4. Multiplication undoes division.

Subtraction is the inverse of addition. For example, start with the number -12. Now add 22 to -12. The result is 10. Next subtract 22 from 10. The result is -12 the original number. Subtracting 22 undoes adding 22. Subtraction undoes addition.

Addition is the inverse of subtraction. For example, start with the number -25. Now subtract -12. The result is -13. Next add -12 to -13. The result is -25, the original number. Adding -13 undoes subtraction of -13. Addition undoes subtraction.

The concept of undoing operations is important when solving equations. The concept of undoing operations works with variables as well as constants. For example, start with the number r. Now add 3 to r. The result is $r + 3$. Next subtract 3 from $r + 3$. The result is $r + 3 - 3$ which equals r.

Balance must be maintained as an equation is solved. When the value on one side of an equation is changed, the value of the other side must be changed by the same amount and operation. For example, if one side of an equation is multiplied by 6, the other side must be multiplied by 6.

Here is an example of solving an equation for the value of a variable.

The equation is $2x = 4$.

Solve.

$2x = 4$

The statement says 2 times x equals 4.

Isolate x, or solve for x, by undoing the multiplication.

$$\frac{2x}{2} = \frac{4}{2}$$

Undo multiplication by using the inverse of multiplication, which is division. Divide the $2x$ by 2. Dividing one side of an equation by 2, requires that the other side be divided by 2 also, dividing both sides by the same number will keep the equation balanced. So, divide the 4 by 2 also.

$x = 2$

The solution is $x = 2$.

The steps as they should appear on paper are shown below.

$$2x = 4$$

$$\frac{2x}{2} = \frac{4}{2}$$

$$x = 2$$

Check the result by replacing x in the <u>original</u> equation with 2, the solution, and see if the equation is true.

$$2x = 4$$

$$2 \cdot 2 = 4$$

$$4 = 4$$

The equation is true. The solution is correct.

Solve the equation below for x. 2x + 3 = -4	The operations performed on the variable x are first, multiplication by 2 and then addition of 3. To solve the equation, isolate the variable by working backwards. Undo the last operation, addition, first.	Undo addition by using the inverse of addition, subtraction. Therefore, subtract 3 from both sides of the equation. Remember to keep the equation balanced. 2x + 3 = -4 2x+ 3 - 3 = -4 - 3	The result is the following equation. 2x = -7

Next, undo the multiplication. Therefore, divide both sides of the equation by 2.
Recall that undoing multiplication requires division. Division is the inverse of multiplication.
 	
Remember to keep the equation balanced.
 	
2x = -7
 	
2x = -7
2   2
	The result is the following equation.
	x = -7
	    2
	
	The equation is solved and x equals
	-7.
	2
		By dividing -7 by 2, x can be written as -3.5.
		
		-7 divided by 2 is the same as -3.5.
		
		Here is the way the solution should appear on paper.
			2x + 3 = -4
			
			2x + 3 - 3 = -4 - 3
			
			2x = -7
			
			2x = -7
			2 2
			
			x = -7 or -3.5
			    2

Check the solution: 2x + 3 = -4 Replace x with -3.5.	2(-3.5) + 3 = -4	-7 + 3 = -4	-4 = -4 **The solution is correct.**
Solve the equation below for f. 3f + 3 = -5 First identify the operations that must be undone.	**SHOW WORK**		**The solution is on the next page.**

Solution

The operations performed on the variable are first multiplication by 3, then addition of 3.

First, undo the addition by using the inverse and subtracting 3 from both sides of the equation. Next, divide both sides of the equation by 3 to undo multiplication.

$$3f + 3 = -5$$

$$3f + 3 - 3 = -5 - 3 \quad \text{Subtract 3 from both sides.}$$

$$3f = -8$$

$$\frac{3f}{3} = \frac{-8}{3} \qquad \text{Divide both side by 3}$$

$$f = \frac{-8}{3}$$

The solution to the equation is $f = \dfrac{-8}{3}$.

Check the solution.

$$3f + 3 = -5$$

$$3 \cdot \frac{-8}{3} + 3 = -5$$

$$-8 + 3 = -5$$

$$-5 = -5$$

The solution is correct.

Solve the equation below for x. $\dfrac{x}{3} - 7 = -5$	The operations performed on the number x are division by 3 and subtraction of 7. Solve the equation by undoing the last operation, subtraction.	Undo subtraction by using the inverse, addition. Therefore, add 7 to <u>both sides</u> of the equation. Remember to keep the equation balanced. $\dfrac{x}{3} - 7 = -5$ $\dfrac{x}{3} - 7 + 7 = -5 + 7$	The result is the following equation. $\dfrac{x}{3} = 2$
Next, undo the division. Undo the division by using multiplication. Therefore, multiply <u>both sides</u> of the equation by 3. Remember to keep the equation balanced. $\dfrac{x}{3} \cdot 3 = 2 \cdot 3$	The result is the following equation. $x = 6$ The equation is solved and x equals 6.	The steps as they should appear on paper are shown next.	$\dfrac{x}{3} - 7 = -5$ $\dfrac{x}{3} - 7 + 7 = -5 + 7$ $\dfrac{x}{3} = 2$ $\dfrac{x}{3} \cdot 3 = 2 \cdot 3$ x = 6
Check the solution. $\dfrac{x}{3} - 7 = -5$ Replace x with 6.	$\dfrac{6}{3} - 7 = -5$	2 - 7 = -5	-5 = -5 **The equation is true.**
Solve the equation below. $\dfrac{x}{5} + 7 = 8$ First identify the operations to undo.	**SHOW WORK**		**The solution is on the next page.**

Solution

The operations performed on the variable are first division by 5 and then addition of 7. First, undo the addition by subtracting 7 from both sides of the equation. Subtracting is the inverse of addition. Next, multiply both sides of the equation by 5 to undo the division.

$$\frac{x}{5} + 7 = 8$$

$$\frac{x}{5} + 7 - 7 = 8 - 7 \quad \text{Subtract 7 from both sides.}$$

$$\frac{x}{5} = 1$$

$$\frac{x}{5} \cdot 5 = 1 \cdot 5 \quad \text{Multiply both sides by 5}$$

$$x = 5$$

The solution to the equation is $x = 5$.

Check the solution.

$$\frac{x}{5} + 7 = 8$$

$$\frac{5}{5} + 7 = 8$$

$$1 + 7 = 8$$

$$8 = 8$$

The solution is correct.

Solve the equation below for p. $11 = \dfrac{p+7}{3}$ When the variable is on the right side of the equal sign switch the left and right side around. In this case, place 11 on the right side and $\dfrac{p+7}{3}$ on the left.	$\dfrac{p+7}{3} = 11$ Now identify the operations that must be undone. Remember that the fraction bar is a grouping symbol. With this in mind, first the 7 is added to p and then the sum is divided by 3.	Remember to undo operations in reverse order. The division is undone first by using the inverse of division, multiplication. Multiply <u>both sides</u> by 3. Remember balance! $\dfrac{p+7}{3} = 11$ $\dfrac{p+7}{3} \cdot 3 = 11 \cdot 3$	The result is the equation $p + 7 = 33$ Next, undo the addition by using its inverse. The inverse of adding is subtracting.
Subtract 7 from <u>both sides</u> of the equation. Remember Balance! $p + 7 = 33$ $p + 7 - 7 = 33 - 7$ The result is $p = 26$	The solution of the equation is 26. **The following is what should appear on paper.**	$11 = \dfrac{p+7}{3}$ $\dfrac{p+7}{3} = 11$ $\dfrac{p+7}{3} \cdot 3 = 11 \cdot 3$	$p + 7 = 33$ $p + 7 - 7 = 33 - 7$ $p = 26$
Check the solution. In the original equation, replace p with 26.	$11 = \dfrac{p+7}{3}$ $11 = \dfrac{26+7}{3}$	$11 = \dfrac{33}{3}$ $11 = 11$	**The solution is correct.**
Solve the equation below. $6 = \dfrac{g-7}{2}$ First identify the operations to undo.	**SHOW WORK**		**The solution is on the next page.**

41

Solution

$$6 = \frac{g - 7}{2}$$

$$\frac{g - 7}{2} = 6$$

$$\frac{g - 7}{2} \cdot 2 = 6 \cdot 2$$

$$g - 7 = 12$$

$$g - 7 + 7 = 12 + 7$$

$$g = 19$$

Check the solution.

$$6 = \frac{g - 7}{2}$$

$$6 = \frac{19 - 7}{2}$$

$$6 = \frac{12}{2}$$

$$6 = 6$$

Correct

A good approach to solving equations is to simplify both sides of the equation first. Simplify the left side of the equal sign, then the right side of the equal sign. Remember to combine like terms and to eliminate parenthesis.

Solve the equation below for r. $2r + 5 - 3r = 7$ First, simplify the left side of the equation. $2r$ and $3r$ are like terms and can be combined.	$2r + 5 - 3r = 7$ can be written as $2r - 3r + 5 = 7$ $2r - 3r = -1r$ or $-r$.	The resulting equation is $-r + 5 = 7$ The left side is simplified. So is the right side.	Next, identify the operations that need to be undone. r is multiplied by -1 and 5 is added on. $-r$ means $-1r$. Undo addition of 5 first. Use the inverse of addition, which is subtraction.
Subtract 5 from both side of the equation. Remember to keep the equation balanced. $-r + 5 = 7$ $-r + 5 - 5 = 7 - 5$	The result is the following equation. $-r = 2$ (Hint: $-r = -1r$) $-r = 2$ means $-1r = 2$ Next, divide <u>both sides</u> by -1 to undo the multiplication.	$\dfrac{-1r}{-1} = \dfrac{2}{-1}$ The result is the following equation. $r = -2$ The equation is solved. **The following is what should appear on paper when solving the equation.**	$2r + 5 - 3r = 7$ $2r - 3r + 5 = 7$ $-r + 5 = 7$ $-r + 5 - 5 = 7 - 5$ $-r = 2$ $\dfrac{-r}{-1} = \dfrac{2}{-1}$ $r = -2$
Check the solution. $2r + 5 - 3r = 7$ Replace r with -2.	$2(-2) + 5 - 3(-2) = 7$ $-4 + 5 + 6 = 7$	$-4 + 5 + 6 = 7$ $1 + 6 = 7$ $7 = 7$	**The solution is correct.**
Solve the equation below. $6p + 7p - 11 = -12$ First, simplify.	**SHOW WORK**		**The solution is on the next page.**

Solution

First, combine like terms on the right. $6p + 7p$ is $13p$. This leaves $13p - 11 = -12$. Identify the operations that need to be undone. p is multiplied by 13 and 11 is subtracted. Undo operations in reverse order, by undoing the subtraction of 11, then undoing the multiplication by 13.

$$6p + 7p - 11 = -12$$

$$13p - 11 = -12$$

$$13p - 11 + 11 = -12 + 11$$

$$13p = -1$$

$$\frac{13p}{13} = \frac{-1}{13}$$

$$p = -\frac{1}{13} \quad \text{is the solution.}$$

Check the solution.

$$6p + 7p - 11 = -12$$

$$6 \cdot -\frac{1}{13} + 7 \cdot -\frac{1}{13} - 11 = -12$$

$$-\frac{6}{13} + -\frac{7}{13} - 11 = -12$$

$$-\frac{13}{13} - 11 = -12$$

$$-1 - 11 = -12$$

$$-12 = -12$$

The solution is correct

Some equations have variables on both sides of the equal sign. By adding to or subtracting from both sides of the equation, this situation can be eliminated.

Here is an example of the situation.

$$3d + 6 = 2d - 7$$

Produce an equation with the variable on one side only by subtracting 2d from both sides. The process is shown below.

$$3d + 6 = 2d - 7$$

$$3d + 6 - 2d = 2d - 7 - 2d$$

$$3d - 2d + 6 = 2d - 2d - 7$$

$$d + 6 = -7$$

Next, proceed to solve the equation by undoing the addition of 6

$$3d + 6 = 2d - 7$$

$$3d + 6 - 2d = 2d - 7 - 2d$$

$$3d - 2d + 6 = 2d - 2d - 7$$

$$d + 6 = -7$$

$$d + 6 - 6 = -7 - 6$$

$d = $ -13 is the solution. **Check the solution.**

Solve the equation below for b. $2(3b+6)=4-2b$ The left side of the equation is not simplified. Parenthesis need to be cleared. First, simplify the left side of the equation by using the distributive property.	$2(3b+6)=4-2b$ $2\cdot 3b+2\cdot 6=4-2b$ $6b+12=4-2b$ Next, get all variables on the left by adding $2b$ to both sides of the equation.	$6b+12=4-2b$ $6b+12+2b=4-2b+2b$ Simplify both sides of the equation. $6b+12+2b=4-2b+2b$ $6b+2b+12=4-2b+2b$ $8b+12=4$	Next, solve the equation by undoing the multiplication by 8 and the addition of 12.
$8b+12=4$ $8b+12-12=4-12$ $8b=-8$ $\dfrac{8b}{8}=\dfrac{-8}{8}$ $b=-1$	**The following is what should appear on paper when solving the equation.**	$2(3b+6)=4-2b$ $2\cdot 3b+2\cdot 6=4-2b$ $6b+12=4-2b$ $6b+12+2b=4-2b+2b$ $6b+2b+12=4$ $8b+12=4$ continue	$8b+12=4$ $8b+12-12=4-12$ $8b=-8$ $\dfrac{8b}{8}=\dfrac{-8}{8}$ $b=-1$
Check the solution. $2(3b+6)=4-2b$	$2(3(-1)+6)=4-2(-1)$ $2(-3+6)=4+2$	$2(3)=4+2$ $6=6$	**The solution is correct.**
Solve. $3+2a=4+3(2a+6)$	**SHOW WORK**		**The solution is on the next page.**

Solution

$$3 + 2a = 4 + 3(2a + 6)$$

$$3 + 2a = 4 + 3 \cdot 2a + 3 \cdot 6$$

$$3 + 2a = 4 + 6a + 18$$

$$3 + 2a = 22 + 6a$$

$$3 + 2a - 6a = 22 + 6a - 6a$$

$$3 - 4a = 22$$

$$-4a + 3 = 22$$

$$-4a + 3 - 3 = 22 - 3$$

$$-4a = 19$$

$$\frac{-4a}{-4} = \frac{19}{-4}$$

$$a = -\frac{19}{4}$$

Check

$$3 + 2a = 4 + 3(2a + 6)$$

$$3 + 2(-\frac{19}{4}) = 4 + 3 \cdot [2(-\frac{19}{4}) + 6]$$

$$3 + -\frac{38}{4} = 4 + 3 \cdot [(-\frac{38}{4}) + 6]$$

$$3 + (-\frac{19}{2}) = 4 + 3 \cdot [(-\frac{19}{2}) + 6]$$

$$3 + (-\frac{19}{2}) = 4 + 3 \cdot [(-\frac{19}{2}) + 6]$$

$$\frac{6}{2} + (-\frac{19}{2}) = 4 + 3 \cdot [(-\frac{19}{2}) + \frac{12}{2}]$$

$$-\frac{13}{2} = 4 + 3(-\frac{7}{2})$$

$$-\frac{13}{2} = \frac{8}{2} + (-\frac{21}{2})$$

$$-\frac{13}{2} = -\frac{13}{2}$$

The solution is $-\dfrac{19}{4}$

Solve the equation below for b. $ab - d = c$ Variables represent numbers. Treat the variables in the same manner as any other number.	Solve $ab - d = c$ for b, by identifying the operations performed on b. First b is multiplied by a, and then d is subtracted from the product. Now solve the equation by undoing the subtraction of d and multiplication by a.	Undo subtraction by using the inverse of subtraction, addition. Add d to both members of the equation. Remember to maintain balance. $ab - d = c$ $ab - d + d = c + d$ The result of adding d to both sides is $ab = c + d.$	Since c and d are not like terms, the two numbers can not be combined. The expression left as $c + d$. Next undo the multiplication by a by using the inverse of multiplication, division. Divide both sides by a.
$ab = c + d$ $\dfrac{ab}{a} = \dfrac{c+d}{a}$ The result of dividing is $b = \dfrac{c+d}{a}$ The expression on the right, $\dfrac{c+d}{a}$, is simplified.	**The complete process that should be shown on paper follows.**	$ab - d = c$ $ab - d + d = c + d$ $ab = c + d$ $\dfrac{ab}{a} = \dfrac{c+d}{a}$ $b = \dfrac{c+d}{a}$	Check the solution by substituting $\dfrac{c+d}{a}$ for b in the original equation and simplifying. $ab - d = c$ $a \cdot \dfrac{c+d}{a} - d = c$ $c + d - d = c$ $c = c$ correct
Note: $a \cdot \dfrac{c+d}{a} = c + d$	Dividing $c + d$ by a then multiplying by a results in the expression $c + d$.	Division and multiplication are inverse operations.	
Solve for B. $A = 2B + 2\pi rh$	**SHOW WORK**		**The solution is on the next page.**

Focus on what is to be accomplished. Isolate B by eliminating $2\pi rh$ from the expression containing B. Determine which operation is needed to eliminate $2\pi rh$. Similarly, the 2 must be eliminated. Determine which operation is needed to do so. B is multiplied by 2, and then $2\pi rh$ is added. Undo the operations in reverse order. Subtract $2\pi rh$ from both sides of the equation then divide both sides by 2.

$$A = 2B + 2\pi rh$$

$$A - 2\pi rh = 2B + 2\pi rh - 2\pi rh$$

$$A - 2\pi rh = 2B$$

$$\frac{A - 2\pi rh}{2} = \frac{2B}{2}$$

$$\frac{A - 2\pi rh}{2} = B$$

$$B = \frac{A - 2\pi rh}{2}$$

Using Formulas

Many problems can be solved by using formulas. Formulas use symbols to show mathematical relationships. For example, the area of a rectangle is related to the base and the height of the rectangle. The symbols A, b and h are used to represent the area, base and height respectively. The formula A = bh shows the mathematical relationship between the three values. The formula relates the value of the area of a rectangle to multiplying the base times the height of the rectangle. See the illustration below.

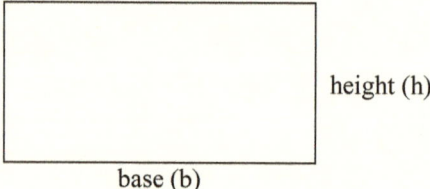

height (h)

base (b)

Below is an example of finding the area of a rectangle given the base and the height.

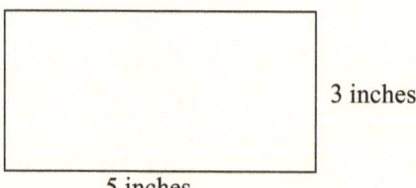

3 inches

5 inches

The base is 5 inches and the height is 3 inches.

Replace the b with 5 and the h with 3 in the formula A = bh.

$A = 5 \cdot 3$ Multiply.

$A = 15$

The area is 15 square inches or in other terms 15 in².

Measurements are reported in units. Since the rectangle units were inches, the units for the area are square inches (in²). 5 inches multiplied by 3 inches is 15 in².

The perimeter of a rectangle can be found when given the base and height. Often with rectangles the terms base and height are replaced with length and width. The longer side is usually considered the length. See the illustration below.

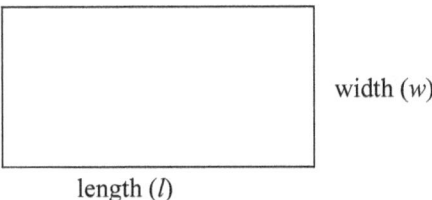

width (w)

length (l)

The perimeter is found by using the formula $P = 2(l) + 2(w)$. The formula says that the perimeter is the sum of twice the length and twice the width of the rectangle. An example of finding the perimeter of a rectangle given the length and width is shown below.

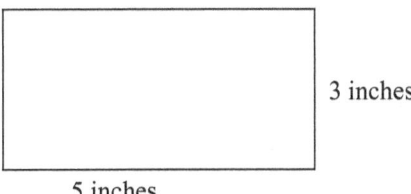

3 inches

5 inches

The length is 5 inches and the width is 3 inches. Replace l with 5 and the w with 3 in the formula $P = 2(l) + 2(w)$.

$$P = 2(5) + 2(3) \qquad \text{Simplify.}$$
$$P = 16$$

The perimeter is 16 inches.

Next is an example of how to use the formula for the area of a rectangle, to find the height given the area and the base. Use the rectangle on the next page.

$$A = bh$$

Area = 20 ft²	h feet

8 feet

Find the value of the height of the rectangle above by using the formula A = bh.

Replace A with 20 and b with 8.

$$A = bh$$
$$20 = 8h$$

Next, solve the equation for h by undoing the multiplication of h. Division is the inverse of multiplication so divide both sides of the equation by 8.

$$20 = 8h$$

$$\frac{20}{8} = \frac{8h}{8}$$

$$2.5 = h$$
or
$$h = 2.5$$

The height of the rectangle is 2.5 feet.

Sometimes it is more convenient to manipulate a formula to isolate a particular variable. Solving an equation for a specified variable will accomplish this. For example, solve the formula for the perimeter of a rectangle, $P = 2(l) + 2(w)$ for l, the length.

Solve $P = 2(l) + 2(w)$ for l.

Solve $P = 2(l) + 2(w)$ for l.

$$P = 2(l) + 2(w)$$

l is multiplied by 2 and then $2w$ is added.

Undo the operations in reverse order as done with any linear equation. First, undo the addition of $2w$ by using the inverse of addition, subtraction. Subtract $2w$ from both sides of the equation. Remember to keep the equation balanced.

$$P - 2(w) = 2(l) + 2(w) - 2(w)$$

$$P - 2(w) = 2(l)$$

Next, use the inverse of multiplication, division. Divide both sides by 2.

$$P - 2(w) = 2(l)$$

$$\frac{P - 2(w)}{2} = \frac{2(l)}{2}$$

$$\frac{P - 2(w)}{2} = l$$

$$l = \frac{P - 2(w)}{2}$$

The altered form of the formula can be useful if there are multiple situations where the perimeter and the width of a rectangle are given and the length is to be found. In each case below, P and w can be replaced to find the value of l.

w = 5 cm

P = 35 cm	P = 42 cm
$l = ?$	$l = ?$

w = 6 cm

An example of using the altered form to find the length of a rectangle is on the next page.

P = 35 cm   w = 5 cm

l =? cm

$$l = \frac{P - 2(w)}{2}$$

Replace P with 34 and w with 5.

$$l = \frac{35 - 2(5)}{2}$$

Simplify by using the order of operations.

$$l = \frac{35 - 10}{2}$$

$$l = \frac{25}{2}$$

$$l = 12.5$$

The length is 12.5 cm.

Converting temperature measurements from degrees Fahrenheit to degrees Celsius is done by using the formula, $C = \frac{5}{9}(F - 32)$, where C is the number of degrees Celsius and F is the number of degrees Fahrenheit.

Convert the boiling point of water, 212^0 Fahrenheit, to degrees Celsius. Use the formula $C = \frac{5}{9}(F - 32)$ by replacing F with 212 and simplifying to find the value of C.

Convert the boiling point of water, 212° Fahrenheit, to degrees Celsius.

$$C = \frac{5}{9}(F - 32)$$

$$C = \frac{5}{9}(212 - 32)$$

$$C = \frac{5}{9}(180)$$

$$C = 100$$

212° Fahrenheit is the same temperature as 100° Celsius.
Both are the boiling point of water.

Converting temperature measurements from degrees Celsius to degrees Fahrenheit is done by using the formula, $F = \frac{9}{5}(C) + 32$, where C is the number of degrees Celsius and F is the number of degrees Fahrenheit. The freezing point of water in degrees Celsius is 0°. Use the formula, $F = \frac{9}{5}(C) + 32$ to do so.

Replace C with 0 and simplify by using the order of operations.

$$F = \frac{9}{5}(0) + 32$$

$$F = 0 + 32$$

$$F = 32$$

0° Celsius is the same as 32° Fahrenheit. Both are the freezing point of water.

A rectangular based prism is shown below. The formula to find the volume is $V = lwh$. The volume is equal to the length multiplied by the width and that product multiplied by the height. The formula for the surface area is $SA = 2lw + 2lh + 2wh$. The surface area is equal to two times the product of the length and width, added to two times the product of the length and height, added to two times the product of the width and height.

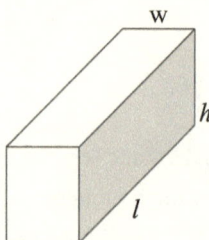

What is the surface area of the rectangular based prism below?

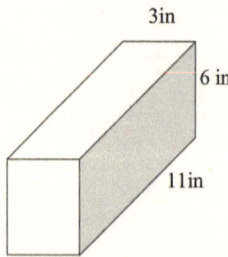

Show the work below. The solution is on the next page.

What is the surface area of the rectangular based prism below?

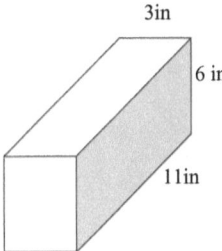

Use the formula for the surface area of a rectangular prism.

$$SA = 2lw + 2lh + 2wh$$

Match the values for l, w and h. $l = 11$, $w = 3$ and $h = 6$
Substitute the values of the variables for the variables.

$$SA = 2(11)(3) + 2(11)(6) + 2(3)(6)$$
$$= 66 + 132 + 36$$
$$= 234$$

The surface area is 234 in².

What is the volume of the rectangular based prism below?

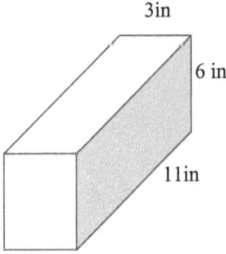

Show the work below. The solution is on the next page.

What is the volume of the rectangular based prism below?

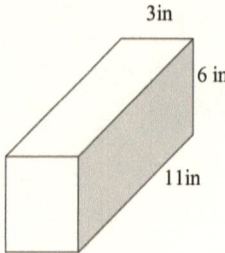

Use the formula for the volume of a rectangular prism.

$$V = lwh$$

Match the values for l, w and h. $l = 11$, $w = 3$ and $h = 6$
Substitute the values of the variables for the variables.

$$V = (11)\,(3)\,(6)$$

$$= 198$$

The volume is 198 in³.

Solving Linear Inequalities Containing One Variable

An inequality is a statement that compares two algebraic expressions that do not represent the same number. The symbols used for inequalities are shown below. Learn these!

$<$ means "is less than"

$\leq$ means "is less than or equal to"

$>$ means "is greater than"

$\geq$ means "is greater than or equal to"

$\neq$ means "is not equal to"

Learn the meaning of the symbols. The symbols are simply short notations for verbal phrases. When the symbol $<$ appears, automatically say "is less than". Do the same for the other symbols.

The statement $2 < 6$ is read "two is less than six." The statement is true since 2 is a smaller number than 6. Remember, when comparing numbers the smaller number is to the left on the number line and the larger is to the right on the number line. The statement $2 > 6$ is read "two is greater than six." The statement is false since 2 is a smaller number than 6.

The statement $y < 5$ is read "y is less than five." y is a variable and represents a set of numbers. There are many numbers that y can be replaced with to make the statement true and other numbers that y can be replaced with to make the statement false. The solution to the inequality is the set of numbers y can be replaced with to make the statement true.

The solution to the inequality y < 5 is the set of all numbers less than 5. In other words, the numbers to the left of 5 on the number line. This includes numbers like 4.99, 4.3, 0, $2\frac{1}{2}$, 2, and -3.

The statement y < 5 can be graphed on a number line to illustrate the numbers that make the statement true.

The following is a blank number line.

The number line below contains the graph of y < 5.

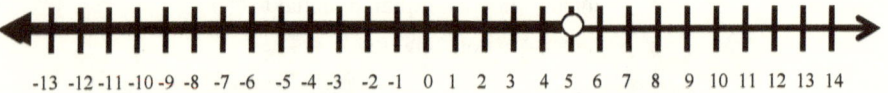

The open circle above the 5 indicates that 5 will not make the statement $y < 5$ true.

5 < 5 is a false statement because 5 is not less than 5. The dark line above the numbers left of 5 indicate that these numbers are solutions of the inequality $y < 5$. The arrow on the left end of the dark line indicates that the solutions don't stop at the end of the number line, but continue forever. In other words numbers like—1,000,000 are included in the solutions. These numbers are solutions because if y is replaced with any of them, the statement $y < 5$ is true. For example, use the number 1, which the dark line lies above. Replace y in the statement $y < 5$ with 1. The statement is now 1 < 5, which is true because 1 is less than 5.

Replacing the < in the statement $y < 5$, with $\leq$, produces a slightly different solution set. The solution set of the new statement $y \leq 5$ includes 5 as well as the numbers less than 5. Remember $\leq$ means less than or equal to. The graph changes to reflect the change in the inequality. The new graph follows.

The following is a blank number line.

Notice the circle above the 5 is now shaded in. This indicates that 5 is a solution of the inequality. When y is replaced with 5 the statement reads $5 \leq 5$, which is true.

Solving an inequality is similar to solving equations. Differences will be identified in the next section. Solving an inequality requires undoing operations to isolate the variable. The symbols <, >, $\leq$ and $\geq$ are symbols that express order. Order must be maintained when solving inequalities. Order is maintained by performing the same operations on both sides of the symbol. For example, if 2 is added to one side of an inequality, then 2 must be added to the other side. This is the same procedure used in solving equations.

Here is an example.

Solve $2y - 7 > 11$.

Similar to solving equations the first step is to identify the operations to be undone. Notice that y is multiplied by 2, then 7 is subtracted. The operations must be undone in

The number line below contains the graph of $y \leq 5$.

reverse order. Undo the subtraction by using the inverse of subtraction, addition. Add 7 to both sides of the inequality to maintain the relationship of order between the two sides of the inequality. The two sides are the left of the symbol, >, and the right of the symbol, >.

Solve

$2y - 7 > 11$

$2y - 7 + 7 > 11 + 7$

$2y > 18$ is the result of adding 7 to both sides.

The inequality is not solved since the y is not yet isolated. Undo the multiplication by using the inverse of multiplication, division. Divide both sides by 2.

$2y > 18$

$$\frac{2y}{2} > \frac{18}{2}$$

$y > 9$ The inequality is solved. The solution is all numbers greater than 9. This does not include 9 since 9 is not greater than 9.

The graph to illustrate the solution is shown below. Notice that the 9 on the number line has an open circle to indicate that 9 is not a solution.

The following is a blank number line.

The number line below contains the graph of $y > 9$. The statement $y > 9$ means all numbers greater than 9.

Solve the inequality below for x. $2x + 3 \leq -4$	The operations performed on the number x are multiplication by 2 and addition of 3. Solve the inequality by undoing the last operation performed, addition.	Undo addition by using the inverse of addition, subtraction. Therefore, subtract 3 from <u>both sides</u> of the inequality. Remember to keep the order in the inequality.	$2x + 3 \leq -4$ $2x + 3 - 3 \leq -4 - 3$ The result is the following inequality. $2x \leq -7$
Next, undo the multiplication. Undo multiplication by dividing. Division is the inverse of multiplication. Therefore, divide <u>both sides</u> of the inequality by 2. Remember to keep the order in the inequality. $2x \leq -7$ $\dfrac{2x}{2} \leq \dfrac{-7}{2}$	The result is the following inequality. $x \leq \dfrac{-7}{2}$ The inequality is solved and x is less than or equal to $\dfrac{-7}{2}.$ The solution to the inequality is all the numbers less than $\dfrac{-7}{2}$ and $\dfrac{-7}{2}.$	Dividing -7 by 2 then x can be expressed as -3.5. -7 divided by 2 is the same as -3.5. **Here is the way the solution should appear on paper.**	$2x + 3 \leq -4$ $2x + 3 - 3 \leq -4 - 3$ $2x \leq -7$ $\dfrac{2x}{3} \leq \dfrac{-7}{2}$ $x \leq \dfrac{-7}{2}$ or $x \leq -3.5$

Graph the solution.

Notice the filled circle on -3.5.

Check the solution. <u>Choose</u> a number from the shaded region of the graph.	-5 is in the shaded region and should satisfy the original inequality	Replace x in $2x + 3 \leq -4$ with -5. $2(-5) + 3 \leq -4$ $-10 + 3 \leq -4$	$-7 \leq -4$ The statement is true. It appears the solution is correct.
Solve the inequality below for f. $3f + 3 > -5$ First identify the operations that must be undone.	SIIOW WORK		**The solution is on the next page.**

Solution

The operations performed on the variable are first multiplication by 3, then addition of 3.

First, undo the addition by using the inverse and subtracting 3 from both sides of the inequality. Next, divide both sides of the inequality by 3 to undo multiplication.

$$3f + 3 > -5$$

$$3f + 3 - 3 > -5 - 3 \quad \text{Subtract 3 from both sides.}$$

$$3f > -8$$

$$\frac{3f}{3} > \frac{-8}{3} \qquad \text{Divide both side by 3}$$

$$f > \frac{-8}{3}$$

The solution to the inequality is $f > \dfrac{-8}{3}$ **or** $f > -2\dfrac{2}{3}$.

Graph the solution.

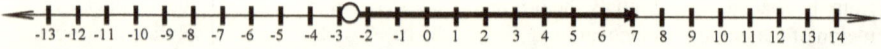

Notice the open circle over $-2\dfrac{2}{3}$. The circle is open because $-2\dfrac{2}{3}$ is not part of the solution set.

Solve the inequality below for x. $\dfrac{x}{3}-7\ge-5$	The operations performed on the number x are division by 3 and subtraction of 7. Solve the inequality by undoing the last operation performed, subtraction.	Undo subtraction using the inverse of subtraction, addition. Therefore, add 7 to <u>both sides</u> of the inequality. Remember to keep the order in the inequality.	$\dfrac{x}{3}-7\ge-5$ $\dfrac{x}{3}-7+7\ge-5+7$ $\dfrac{x}{3}\ge 2$
Next, undo the division by multiplying. Multiplication is the inverse of division. Therefore, multiply <u>both sides</u> of the inequality by 3. Remember to keep the order in the inequality. $\dfrac{x}{3}\ge 2$ $\dfrac{x}{3}\cdot 3\ge 2\cdot 3$	The result is the following inequality. $x\ge 6$ The inequality is solved and x is greater than or equal to 6. The solution to the inequality is all the numbers greater than 6 and 6 itself.	**Next is the way the solution should appear on paper.**	$\dfrac{x}{3}-7\ge-5$ $\dfrac{x}{3}-7+7\ge-5+7$ $\dfrac{x}{3}\ge 2$ $\dfrac{x}{3}\cdot 3\ge 2\cdot 3$ $x\ge 6$
Graph the solution.			

Notice the filled circle on 6.

Check the solution. <u>Choose</u> a number from the region not shaded of the graph.	-5 is in the region not shaded and should **not** satisfy the original inequality	Replace x in $\dfrac{x}{3}-7\ge-5$ with -5. $\dfrac{-5}{3}-7\ge-5$ $-6\dfrac{2}{3}\ge-5$	The statement is false. According to our graph the statement should be false since the number we chose was not shaded. It appears the solution is correct.
Solve the inequality below for f. $\dfrac{x}{5}+7\le 8$ First, identify the operations that must be undone.	**SHOW WORK**		**The solution is on the next page.**

Solution

The operations performed on the variable are first division by 5, then addition of 7. First, undo the addition by subtracting 7 from both sides of the inequality. Subtracting is the inverse of addition. Next, multiply both sides of the inequality by 5 to undo the division.

$$\frac{x}{5} + 7 \leq 8$$

$$\frac{x}{5} + 7 - 7 \leq 8 - 7 \quad \text{Subtract 7 from both sides.}$$

$$\frac{x}{5} \leq 1$$

$$\frac{x}{5} \cdot 5 \leq 1 \cdot 5 \quad \text{Multiply both sides by 5.}$$

$$x \leq 5$$

The solution to the inequality is $x \leq 5$.

Graph the solution.

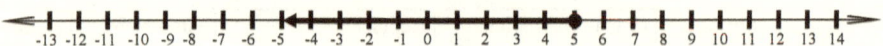

Notice the closed circle on 5. That is because x can equal 5.

$5 \leq 5$ is a true statement.

Check the solution before proceeding.

Solving Linear Inequalities That Require Multiplying or Dividing Both Sides of the Inequality by a Negative

There are situations where solving an inequality is different than solving an equation. The following examples will illustrate these differences. Pay particular attention to solving inequalities that require dividing both sides by a **negative** number or multiplying both sides by a **negative** number. Multiplying or dividing both sides of an inequality by a negative number, affects the order. Here is an example of the affect. Consider the inequality $1 < 2$. Multiply both sides of the inequality by -1. The result of multiplying 1 by -1 is -1. The result of multiplying 2 by -1 is -2. The left side of the inequality is now -1 and the right side is -2. Which is greater, -1 or -2? -1 is greater than -2. The new inequality is $-1 > -2$. The order has been reversed. $1 < 2$ but $-1 > -2$

The number line is good for illustrating this change in order.
Look at the position of 1 with respect to 2 and -1 with respect to -2.

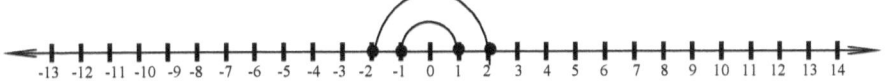

$1 < 2$ but $-1 > -2$

Here is an example of an inequality requiring a change in order to reach the solution.

$$-2u < 4$$

$-2u < 4$ u is multiplied by -2. Therefore divide both sides by -2.

$\dfrac{-2u}{-2} > \dfrac{-4}{-2}$ When dividing both sides of an inequality by a negative, change the order.

$u > 2$ The solution is all numbers greater than 2.

Solve the inequality below for x. $-2x + 3 \leq -4$	The operations performed on the number x are multiplication by -2 and addition of 3. Solve the inequality by undoing the last operation performed, addition.	Undo addition using the inverse of addition, subtraction. Therefore, subtract 3 from <u>both sides</u> of the inequality. Remember to keep the order in the inequality.	$-2x + 3 \leq -4$ $-2x + 3 - 3 \leq -4 - 3$ The result is the following inequality. $-2x \leq -7$
Next, undo the multiplication by dividing. Multiplication's inverse is division. Therefore, divide <u>both sides</u> of the inequality by -2. Remember to **change** the order since both sides are divided by a **Negative**. $-2x \leq -7$ $\dfrac{2x}{-2} \geq \dfrac{-7}{-2}$	The result is the following inequality. $x \geq \dfrac{7}{2}$ The inequality is solved and x is greater than or equal to $\dfrac{7}{2}.$ The solution to the inequality is all the numbers greater than $\dfrac{7}{2}$ and $\dfrac{7}{2}$ itself.	Dividing 7 by 2 to express $\dfrac{7}{2}$ as 3.5. 7 divided by 2 is the same as 3.5. **Here is the way the solution should appear on paper**.	$-2x + 3 \leq -4$ $-2x + 3 - 3 \leq -4 - 3$ $-2x \leq -7$ $\dfrac{-2x}{-2} \geq \dfrac{-7}{-2}$ $x \geq \dfrac{7}{2}$ or $x \geq 3.5$

Graph the solution.

Notice the filled circle on 3.5.

Check the solution. <u>Choose</u> a number from the shaded region of the graph.	5 is in the shaded region and should satisfy the original inequality	Replace x in $-2x + 3 \leq -4$ with 5. $-2(5) + 3 \leq -4$ $-10 + 3 \leq -4$	$-7 \leq -4$ The statement is true. It appears the solution is correct.
Solve the inequality below for f. $-3f + 3 > -9$ First, identify the operations that must be undone.	**SHOW WORK**		**The solution is on the next page.**

Solve the inequality

-3f + 3 > -9   *f* is multiplied by -3 and 3 is added.

Undo addition first.

-3f + 3 - 3 > -9 - 3

-3f > -12   Next undo the multiplication.

$\dfrac{-3f}{-3} < \dfrac{-12}{-3}$ Don't forget to change the order when dividing both sides

by a negative. f < 4

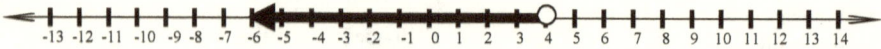

Notice the open circle on 4. 4 isn't a solution for the inequality since 4 is not less than 4.

4 < 4 is a false statement.

Solve the inequality below for f. $-\dfrac{r}{3} \le 11$ First identify the operations that must be undone.	SHOW WORK		The solution is on the next page.

70

Solve the inequality $-\dfrac{r}{3} \le 11$.

$-\dfrac{r}{3} \le 11$ r is divided by -3.

$-3\left(-\dfrac{r}{3}\right) \ge -3\,(11)$ Multiply both sides by -3.

                                        Notice the change in order.

$r \ge -33$

Graphing Relations

A relation is simply a set of ordered pairs. Ordered pairs are written in the form (x, y). For example (3, 5) is an ordered pair where x =3 and y =5. The order of the numbers is important. The ordered pair (5, 3) is not the same as the ordered pair (3, 5). (5, 3) says x = 5 and y = 3. Relations can be illustrated by sets such as the following.

$$\{(1,9), (3,6), (4,8), (11,6)\}$$

Each element of the set is an ordered pair. (1,9) is an ordered pair, (3,6) is an ordered pair, (4,8) is an ordered pair and (11,6) is an ordered pair.

A relation can also be illustrated by writing an equation that states a rule for the relation. An example of such a rule is $f(x) = 2x + 3$ or $y = 2x + 3$. The two equations say the same thing. The notation $f(x)$ is read "f of x." The $f(x)$ is just another way of saying y. The rule states that for each number x, multiply by 2 and add 3 to the product and the result is the value of $f(x)$ or y. The relation is the set of ordered pairs that make the equation true. The ordered pairs are written in the form (x, f(x)) or (x, y). An ordered pair that makes the equation $f(x) = 2x +3$ is (4, 11) where x = 4 and $f(x) = 11$. The reason that it makes the equation true is replacing x with 4 and f(x) with 11 results in the statement $11 = 2(4) + 3$. This is true since $2(4) + 3$ is 11. The ordered pair (4, 11) is merely a tiny part of the relation with rule $f(x) = 2x + 3$. There are many ordered pairs in the set. Listing all the ordered pairs is not possible. However, illustrating the relation with a graph is possible. This provides a visual representation (picture) of the relation. The picture will be graphed on a coordinate plane as shown below. Each ordered pair matches with a point in the plane. The first number in the ordered pair or first coordinate, tells how far to move horizontally from the origin. The origin, shown on the graph, has the ordered pair (0, 0). The second coordinate tells how far to move vertically.

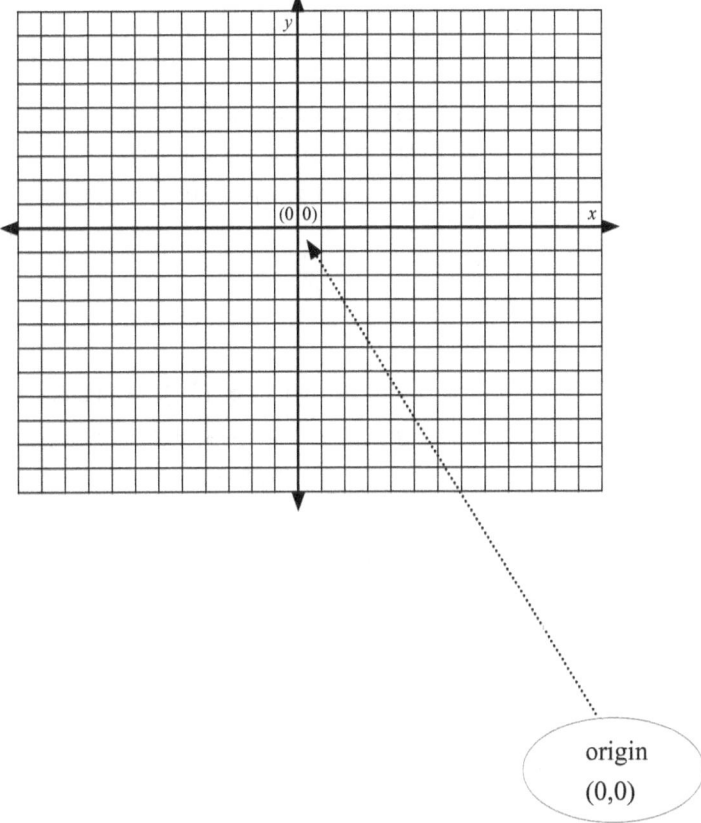

origin
(0,0)

The next page shows several points graphed with instruction on the procedure to graph.

73

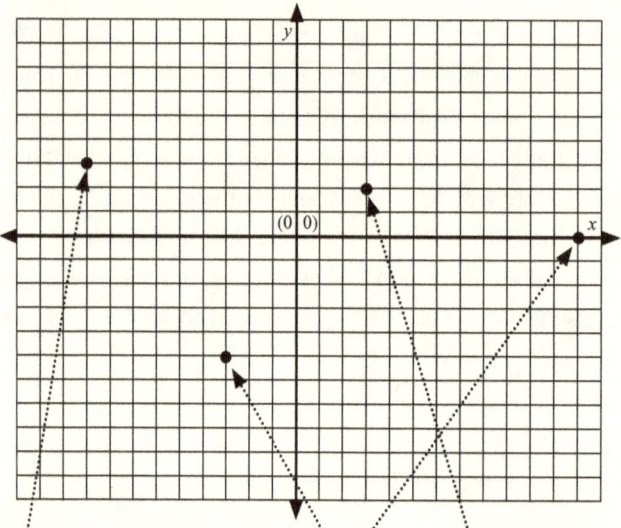

(-9,3) Place pointer on (0,0), count nine left and 3 up and make a small dot. The dot on the graph is much larger than needed.

(3,2) Place pointer on (0,0), count three right and 2 up and make a small dot. The dot on the graph is much larger than needed.

(12,0) Place pointer on (0,0), count twelve right and 0 up and make a small dot. The dot on the graph is much larger than needed.

(-3,-5) Place pointer on (0,0), count three left and 5 down and make a small dot. The dot on the graph is much larger than needed.

Graph this relation by generating ordered pairs that satisfy the relation. Generate these by using a table of values for one variable and calculate the values of the second variable. Shown earlier was the ordered pair (4, 11) as part of the relation f(x) = 2x + 3. Below is a method to find others.

1. Create a table as shown below. This will be used to generate ordered pairs that are a part of the relation.

x	f(x)=2x+3	f(x)	(x, f(x))

2. Choose a value for x, say -1. Place -1 in the column under x.

x	f(x)=2x+3	f(x)	(x, f(x))
-1			

3. Next, replace x in the relation with -1.

x	f(x)=2x+3	f(x)	(x, f(x))
-1	f(-1)=2(-1)+3		

4. Find the value of f(-1) by simplifying 2(-1)+3. 2(-1)+3 = -2 + 3 = 1

x	f(x)=2x+3	f(x)	(x, f(x))
-1	f(-1)=2(-1)+3	1	

5. Write the ordered pair generated.

x	f(x)=2x+3	f(x)	(x, f(x))
-1	f(-1)=2(-1)+3	1	(-1, 1)

6. ***Choose*** more values for x to find more ordered pairs.

x	f(x)=2x+3	f(x)	(x, f(x))
-1	f(-1)=2(-1)+3	1	(-1, 1)
0	f(0)=2(0)+3	3	(0, 3)
1	f(1)=2(1)+3	5	(1, 5)
2	f(2)=2(2)+3	7	(2, 7)
3	f(3)=2(3)+3	9	(3, 9)

The graph contains arrows pointing to the points generated from the table.

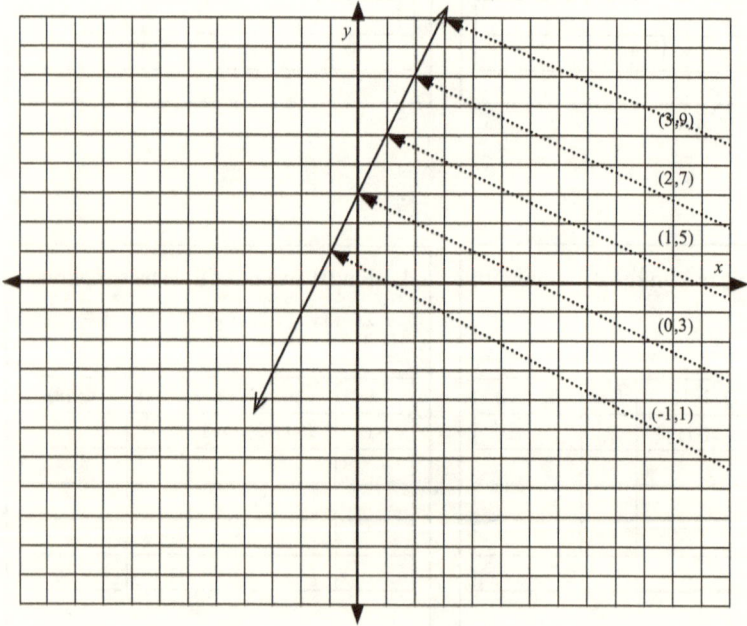

The graph appears on the next page without the arrows pointing to the generated points.

$$f(x) = 2x + 3$$

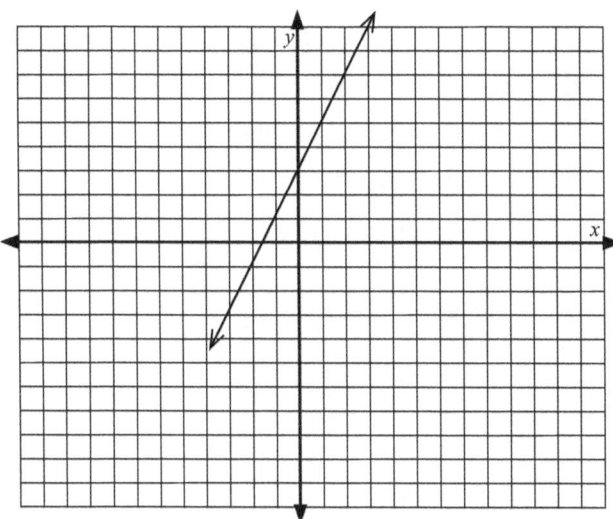

Graph the relation $y = \frac{1}{2}x - 5$.

Remember that a relation is a set of ordered pairs. Graph the relation above by finding ordered pairs that will satisfy the equation. In other words, find ordered pairs that make the equation true. **Choose** values for x and calculate the corresponding values for y.

x	$y = \frac{1}{2}x - 5$	y	(x, y)
-2	$y = \frac{1}{2}(-2) - 5$	-6	(-2, -6)
0	$y = \frac{1}{2}(0) - 5$	-5	(0, -5)
2	$y = \frac{1}{2}(2) - 5$	-4	(2, -4)
4	$y = \frac{1}{2}(4) - 5$	-3	(4, -3)
6	$y = \frac{1}{2}(6) - 5$	-2	(6, -2)

Place small dots on the graph for each point generated. The dots below are far larger than needed.

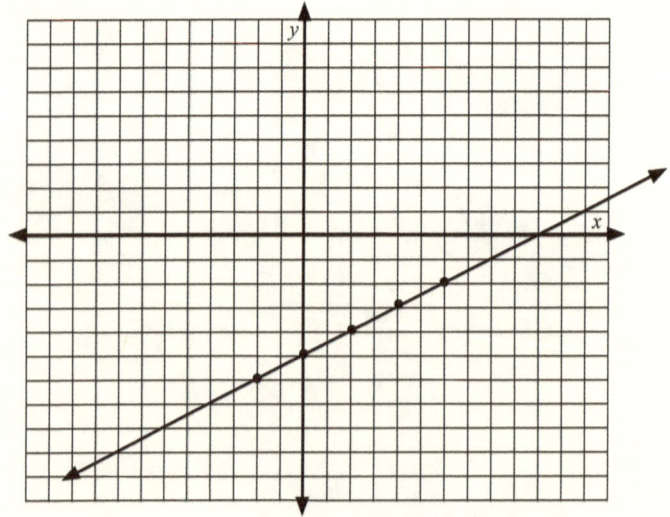

The graph without the dots is on the next page.

$$y = \frac{1}{2}x - 5$$

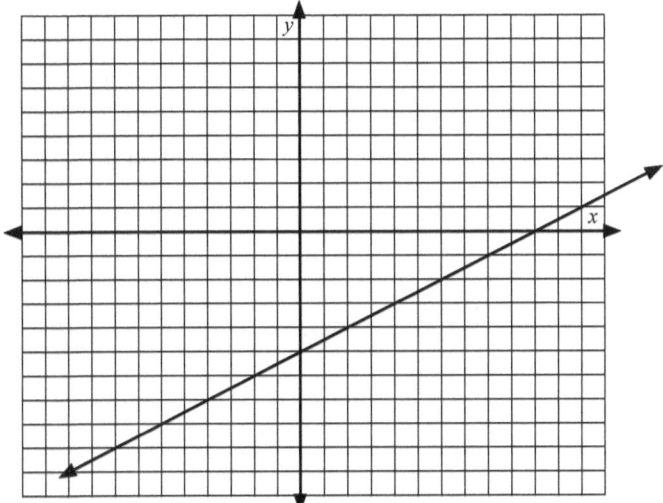

Graph the relation $f(x) = x^2$.

Remember that a relation is a set of ordered pairs. Graph the relation above by generating ordered pairs that will satisfy the equation. In other words, find ordered pairs that make the equation true. **Choose** values for x and calculate the corresponding values for y.

x	$f(x) = x^2$	f(x)	(x, f(x))
-2	$f(-2) = (-2)^2$	4	(-2,4)
-1	$f(-1) = (-1)^2$	1	(-1,1)
0	$f(0) = 0^2$	0	(0,0)
1	$f(1) = 1^2$	1	(1,1)
3	$f(3) = 3^2$	9	(3,9)

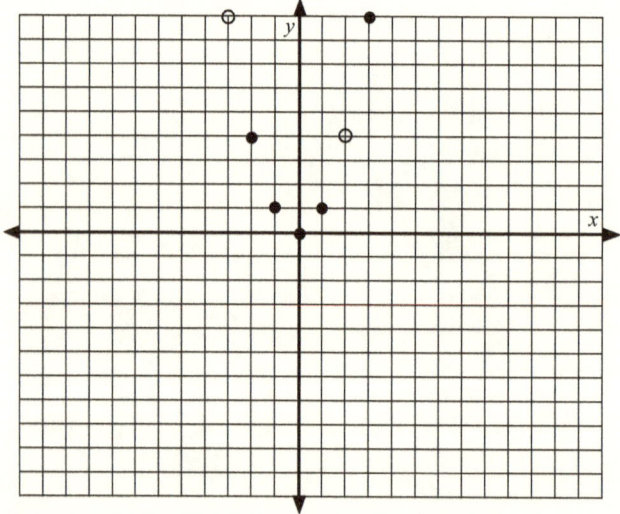

Notice that there are two points that were not on the table. The points have been inserted to give a better picture of the relation. The two ordered pairs (-3,9) and (2,4) will satisfy the rule for the relation. Check to make certain they satisfy the rule! Connect the points to form a smooth curve similar to a U on the graph below. Start with the upper left point, move down to the next point on the right and continue until reaching the upper right point. The curve is just a small part of the relation. Show that the relation continues by placing arrows at the ends.

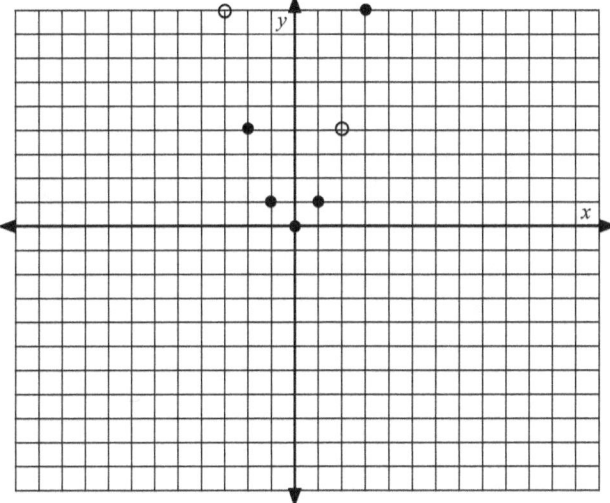

An illustration of how the relation, $f(x) = x^2$ should look is shown below. If it does not, correct it.

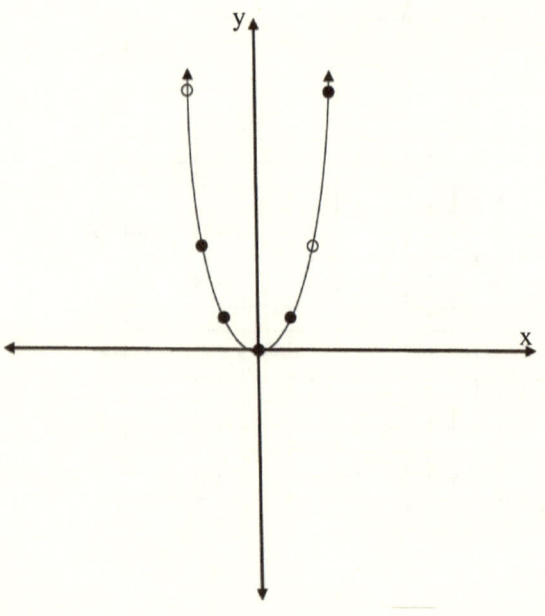

Graph the relation $f(x) = -2x + \frac{2}{3}$. Recall that a relation is a set of ordered pairs. Graph the relation above by finding ordered pairs that will satisfy the equation. In other words, find ordered pairs that make the equation true. **Choose** values for x and calculate the corresponding values for y.

x	$f(x) = -2x + \frac{2}{3}$	y	(x, y)
-3	$f(x) = -2(-3) + \frac{2}{3}$	$6\frac{2}{3}$	$(-3, 6\frac{2}{3})$
-2	$f(x) = -2(-2) + \frac{2}{3}$	$4\frac{2}{3}$	$(-2, 4\frac{2}{3})$
0	$f(x) = -2(0) + \frac{2}{3}$	$\frac{2}{3}$	$(0, \frac{2}{3})$
2	$f(x) = -2(2) + \frac{2}{3}$	$-3\frac{1}{3}$	$(2, -3\frac{1}{3})$
3	$f(x) = -2(3) + \frac{2}{3}$	$-5\frac{1}{3}$	$(3, -5\frac{1}{3})$

Place small dots on the graph for each point. The dots below are larger than needed.

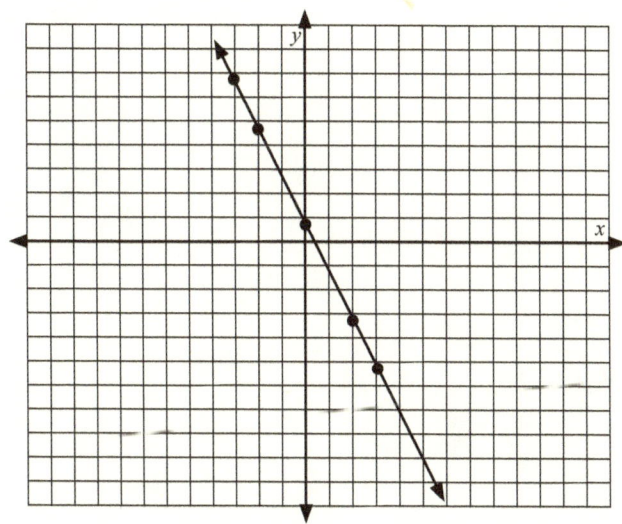

Practice locating the points on the graph. For example, locate $(-2, 4\frac{2}{3})$ by placing the pointer on (0,0) and counting left two and up $4\frac{2}{3}$, then make the dot. The graph without the dots is on the next page.

$$f(x) = -2x + \frac{2}{3}$$

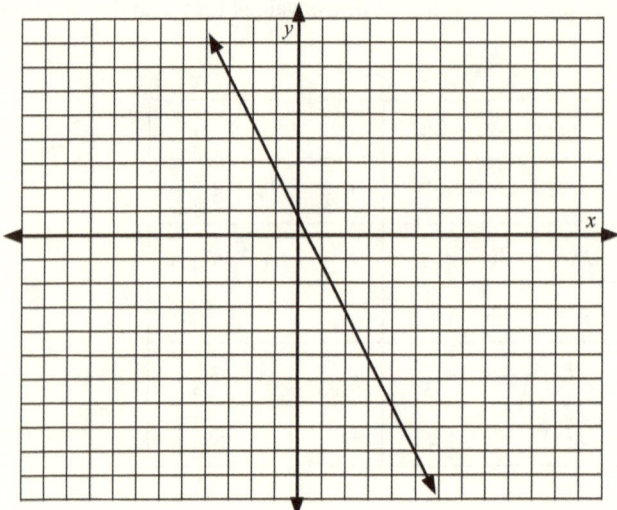

Graph the relation $f(x) = 2x^2 - 3$.

Remember that a relation is a set of ordered pairs. Graph the relation above by finding ordered pairs that will satisfy the equation. In other words, find ordered pairs that make the equation true. **Choose** values for x and calculate the corresponding values for y.

x	$f(x) = 2x^2 - 3$	f(x)	(x, f(x))
-2	$f(x) = 2(-2)^2 - 3$	5	(-2,5)
-1	$f(x) = 2(-1)^2 - 3$	-1	(-1,-1)
0	$f(x) = 2 \cdot 0^2 - 3$	-3	(0,-3)
1	$f(x) = 2 \cdot 1^2 - 3$	-1	(1,-1)
3	$f(x) = 2 \cdot 3^2 - 3$	15	(3,15)

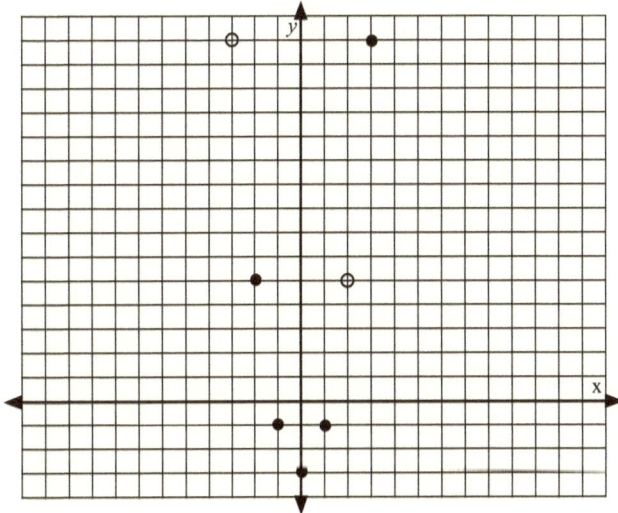

Notice that there are two points that were not on the table. The points are inserted to give a better picture of the relation. The two ordered pairs (-3,15) and (2,5) will satisfy the rule for the relation. Check to make certain they satisfy the rule! Connect the points to form a smooth curve similar to a U on the graph below. Start with the upper left point move down to the next point on the right and continue until reaching the upper right point. The curve is just a small part of the relation. Show that the relation continues by placing arrows at the ends.

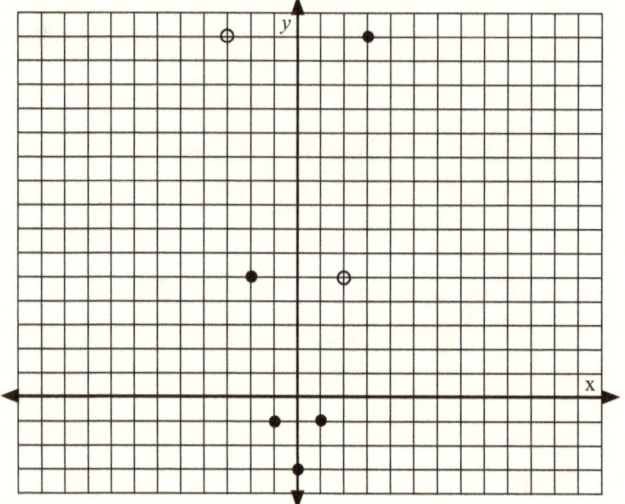

An illustration of how the relation, $f(x) = 2x^2 - 3$ should appear is shown below. If it does not, correct it.

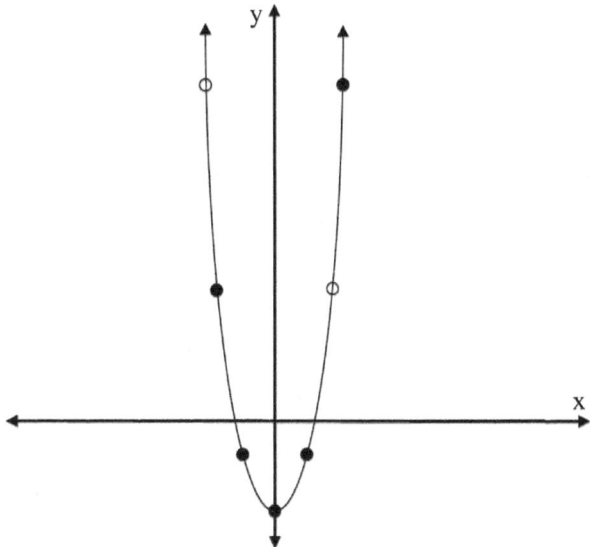

Linear Equations in Two Variables

An equation is a statement that contains an equal sign stating that two expressions represent or equal the same number. Below are examples of equations.

$$2w^2 + 2y = 4$$ quadratic in two variables (not linear)

$$2w + 3z = -7$$ linear in two variables

$$3r - 5s = 17$$ linear in two variables

$$11v = 165n^3$$ cubic in two variables (not linear)

$$2n - 7 = 21 - m$$ linear in two variables

$$-\frac{t}{5} - 11u = 21$$ linear in two variables

$$k^2 + 3 = 12$$ quadratic in one variable (not linear)

$$3x - 2y + z = 11$$ linear in three variables

The equations above are called open sentences because one or more of the numbers are represented by a variable. When two variables appear in the equation and the variables are raised to the first power (no exponent represents exponent 1), the equation is called a linear equation in two variables. There are three common forms for writing these equations.

$$Ax + By = C \text{ Standard Form}$$
$$y - y_1 = m(x - x_1) \text{ Point-Slope Form}$$
$$y = mx + b \text{ Slope Intercept Form}$$

The three forms will be examined individually later in the text. Linear equations in two variables are rules for relations. They represent a set of ordered pairs. Consider the equation, in slope-intercept form, $y = 3x + 7$. The rule says for every value substituted for x, find the value of y by using the following process. Multiply the value of x by 3, and then add 7 to this product. Follow the example below.

Substitute the value 5 for x in the equation $y = 3x + 7$.
$$y = 3 \cdot 5 + 7$$
$$y = 15 + 7$$
$$y = 22$$

The result is an ordered pair, (5, 22).

Slope-intercept form, $y = mx + b$, is commonly used because one of the two variables is isolated. The form allows a quick format to substitute for one variable and solve for the other. Refer to the example above. The value 5 was substituted for x and the corresponding value for y, 22, was readily found. The variable x can be replaced with any value. Then, the corresponding value for y can be found by substituting for x. If many points produced from the equation were graphed, they would all lie on the same line. **The graph of a linear equation in two variables is a line.** An example of the graph of an equation is below and on the following page.

Graph $y = 2x - 1$
By replacing x with 0, y is -1. If x is 1, y is 1. If x is 2, y is 3.
The ordered pairs are (0, -1), (1, 1) and (2, 3).
The graph with these points highlighted is on the next page.

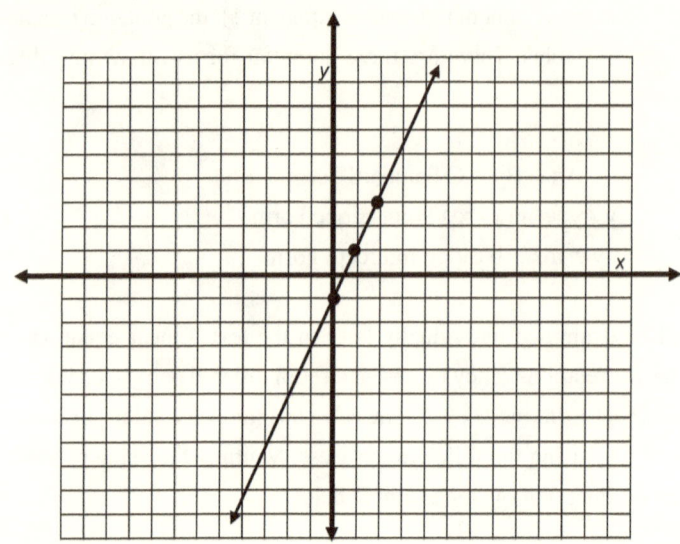

The line, as any line except a vertical line, has a **slope**. The slope assigns a value to the steepness of a line. The slope compares the vertical and horizontal difference that must be covered, to move from one point on the graph, to another. The comparison is expressed as the ratio of two numbers. Consider the line below with two points high-lighted. When starting with the lower point and moving to the higher, the slope is found to be $\frac{2}{1}$ or 2. By starting at the point (1, 1) and rising a sufficient amount to be in line horizontally with (2, 3), a distance of 2 units is covered. After moving vertically, a

horizontal distance of 1 unit is covered to reach the point (2, 3).

Starting with the higher point will produce the same slope. By starting with the higher point (use pencil) and moving down to the lower, a vertical distance of -2 is covered. By moving left to reach the lower point a horizontal distance of -1 is covered. $\frac{2}{1} = \frac{-2}{-1} = 2$

Slopes of lines in four situations

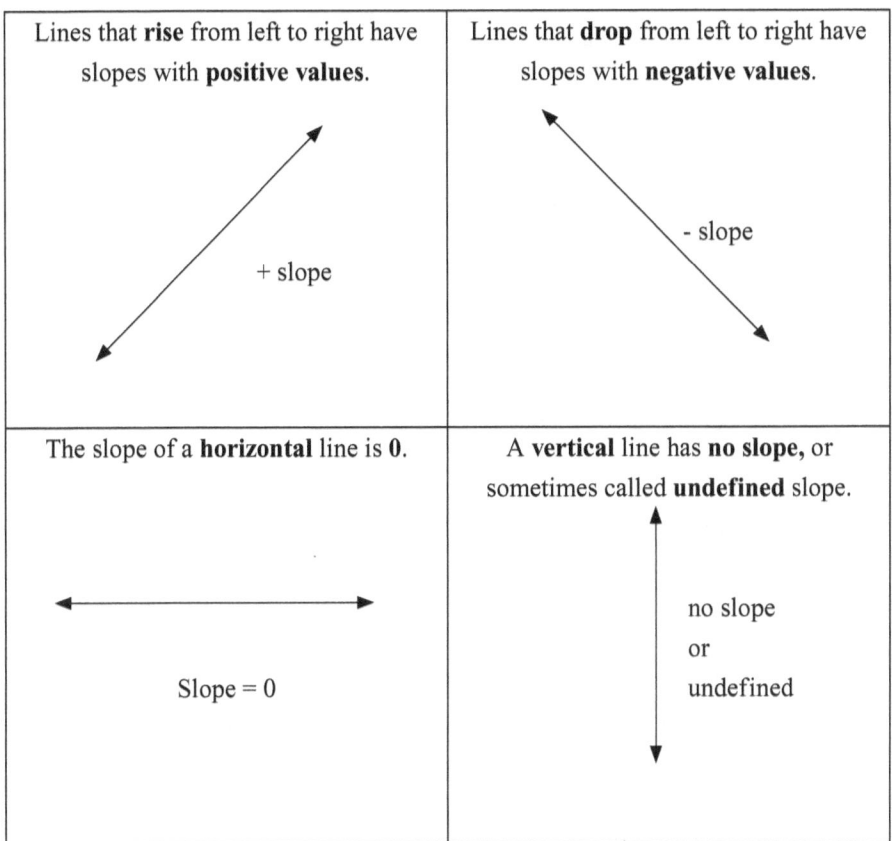

Lines that **rise** from left to right have slopes with **positive values**.	Lines that **drop** from left to right have slopes with **negative values**.
+ slope	- slope
The slope of a **horizontal** line is **0**.	A **vertical** line has **no slope,** or sometimes called **undefined** slope.
Slope = 0	no slope or undefined

The slope of a horizontal line is zero. The steeper the line, the farther the value of the slope is from zero. A line with slope $\frac{1}{2}$ is less steep then a line with slope 2. A line with slope $\frac{1}{2}$ is less steep then a line with slope -2. A line with slope 2 has the same steepness as a line with slope -2. The only difference is that one rises from left to right while the other drops from left to right.

The three forms for the equation of a line will be discussed in the following pages.

Slope-intercept form

The equation of a line can be written in the form $y = mx + b$.

The m represents the slope of the line and b represents the y-intercept of the line.

The y-intercept is the point where a line intersects the y-axis. See the illustration below.

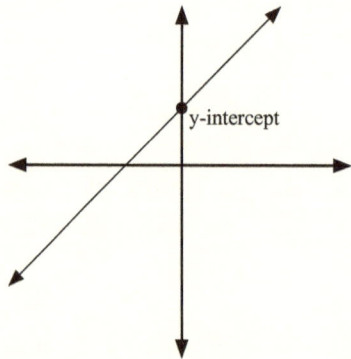

The graph of the line with equation $y = 4x + 2$ has slope 4 or $\frac{4}{1}$ and y-intercept 2. The line crosses the y-axis at the point (0,2). Since the y-intercept is on the y-axis, the value of x is **always** 0 at the y-intercept.

The graph of the line with equation $y = -\frac{2}{3}x - 5$ has slope $-\frac{2}{3}$ and y-intercept -5. The line crosses the y-axis at the point (0,-5). Since the y-intercept is on the y-axis, the value of x is **always** 0 at the y-intercept.

Given the equation of a line in slope-intercept form, graphing is a straight forward process. Examples of graphing using the slope and y-intercept are on the next page.

Graph the line with equation y = 4x +2.

The slope, again, is 4 or $\frac{4}{1}$ and the y-intercept is 2.

First, plot the y-intercept, (0,2).

Use the slope by rising 4 from the y-intercept.

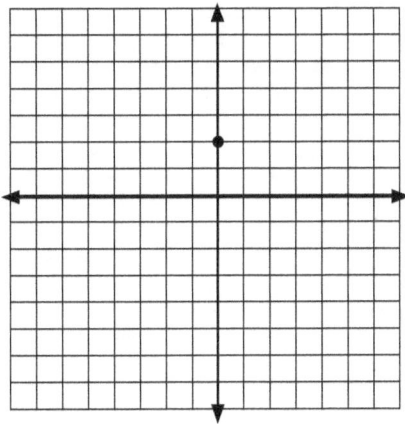

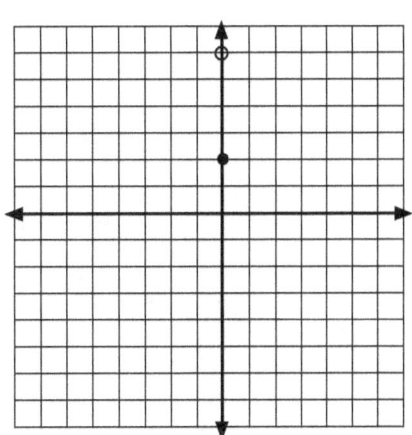

Next, run 1 right. This is a second point on the line.

Draw the line through the points.

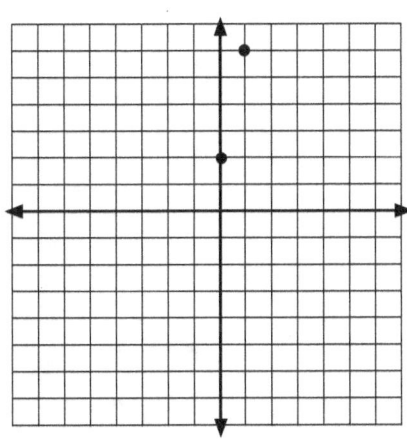

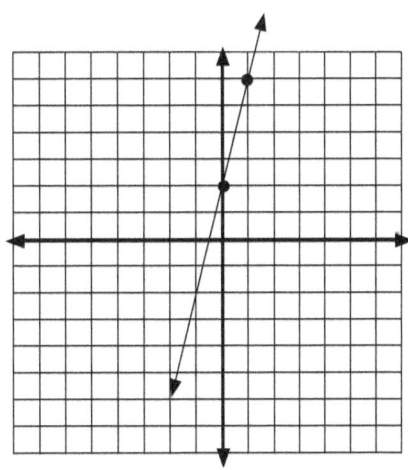

Graph the line with equation $y = -\frac{2}{3}x - 5$. ($y = -\frac{2}{3}x + (-5)$)

The slope, again, is $-\frac{2}{3}$ and the y-intercept is -5.

First, plot the y-intercept, (0,-5).

Use the slope by rising -2 from the y-intercept.

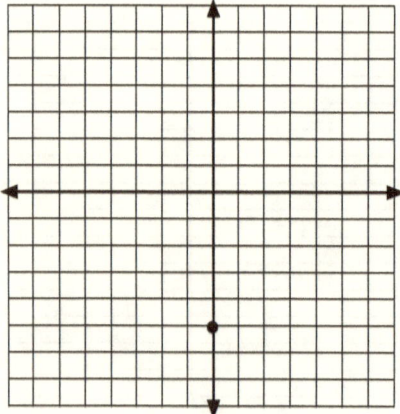

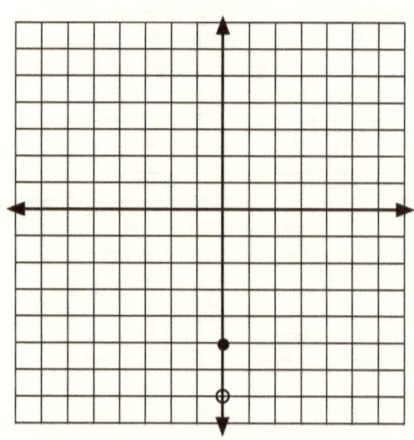

Next, run 3 right. This is a second point on the line.

Draw the line through the points.

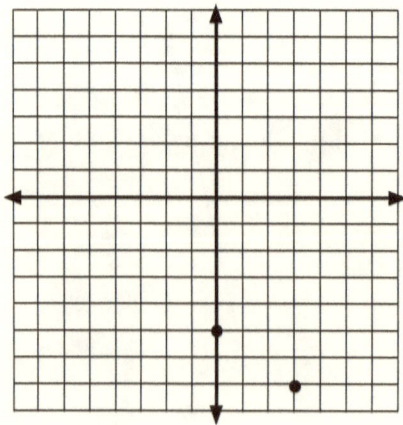

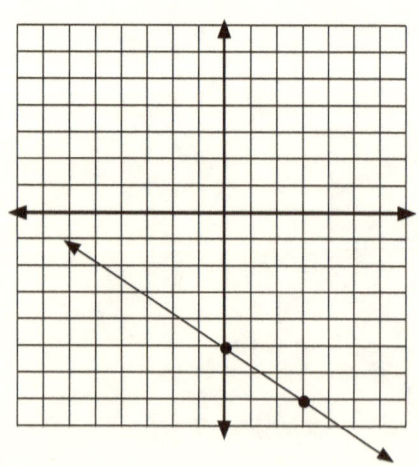

Point-slope form

The equation of a line can be written in the form
$$y - y_1 = m(x - x_1)$$
The m represents the slope of the line and a point on the line is (x_1, y_1)
The point (x_1, y_1) can be any point on the line. See the illustration below.

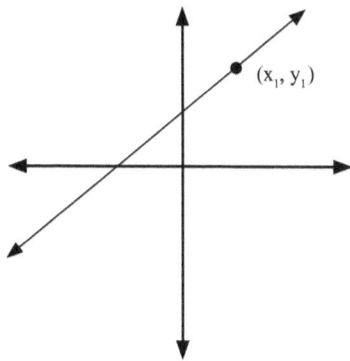

Match the equation $y - 2 = \frac{1}{4}(x-1)$ with the equation $y - y_1 = m(x - x_1)$. $x_1 = 1$, $y_1 = 2$ and $m = \frac{1}{4}$. The graph of the line with equation $y - 2 = \frac{1}{4}(x-1)$ has slope $\frac{1}{4}$ and point $(1, 2)$.

The line goes through the point $(1, 2)$. Plot the given point, $(1, 2)$ first. Now, move up 1 and right 4 to find a second point. Next, draw the line.

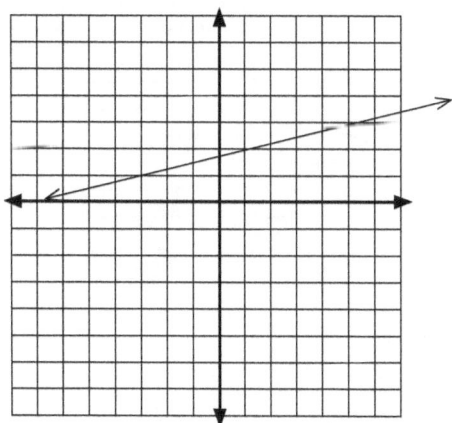

The graph of the line with equation $y + 2 = -3(x + 1)$ has slope -3 or $-\frac{3}{1}$ and passes through the point (-1, -2).

$$y + 2 = -3(x + 1)$$

can be written as

$$y - (-2) = -3(x - (-1))$$

Point slope form is written with subtraction signs after the x and y.

$$y - y_1 = m(x - x_1)$$

To identify the point, the equation must follow the same form.

To graph the line with equation $y + 2 = -3(x + 1)$, plot the given point, (-1, -2) first. Next, move down 3 and right 1 to find a second point.

Finally, draw the line through the points

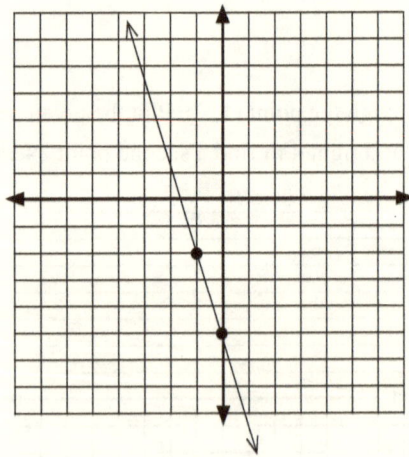

Standard form

The equation of a line can be written in the form

Ax + By = C.

A, B and C are constants, but do not represent the slope or a point individually. The equation can be manipulated, by using the properties of equality, to find the slope and the y-intercept.

To do so, solve the equation, Ax + By = C for y. y is multiplied by B and Ax is added. Undo the operations in reverse order.

Subtract Ax from both sides of the equation first. Then, divide both sides by B.

$$Ax - Ax + By = C - Ax$$

$$By = -Ax + C$$

$$\frac{B}{B}y = -\frac{A}{B}x + \frac{C}{B}$$

$$y = -\frac{A}{B}x + \frac{C}{B}$$

Match the resulting equation with slope intercept form, y = mx + b.

$$m = -\frac{A}{B} \text{ and } b = \frac{C}{B}$$

With these relationships the slope and y-intercept of the line with equation 3x + 2y = 11 can be found.

First, identify A, B and C by matching the equation 3x + 2y = 11 with Ax + By = C.

$$A = 3, B = 2 \text{ and } C = 11 \quad m = -\frac{A}{B} = -\frac{3}{2} \text{ and } b = \frac{C}{B} = \frac{11}{2}$$

The slope is $-\dfrac{3}{2}$ and the y-intercept is $\dfrac{11}{2}$. The graph of the line is shown below.

$$3x + 2y = 11$$

The slope is $-\dfrac{3}{2}$ and the y-intercept is $\dfrac{11}{2}$ or 5.5.

Plot $\left(0, \dfrac{11}{2}\right)$ first, then move down 3 and right 2 to find a second point. Draw the line.

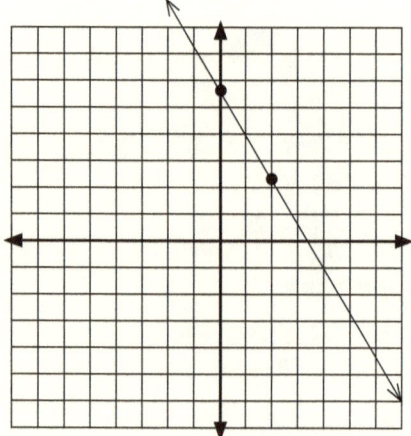

The line with equation $-4x + 3y = -11$ has the following slope and y-intercept.

$$-4x + 3y = -11$$

$$A = -4, B = 3 \text{ and } C = -11$$

$$m = -\frac{A}{B} = -\frac{-4}{3} = \frac{4}{3} \text{ and } b = \frac{C}{B} = \frac{-11}{3} = -3\frac{2}{3}$$

The graph of the line is on the next page

Graph -4x + 3y = -11

The slope is $\frac{4}{3}$ and the y-intercept is $-3\frac{2}{3}$

Plot the point $\left(0, -3\frac{2}{3}\right)$ first, and then move up 4 right 3 to find a second point.
Draw the line.

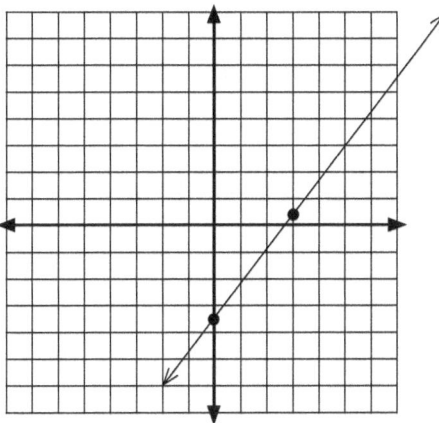

Writing the Equation of a Line

An equation for a line can be written in slope-intercept form or point-slope form when the right information about the line is given. The slope and the y-intercept must be known or produced to write an equation of a line in slope-intercept form. The form $y = mx + b$ requires that only the slope, m, and the y-intercept, b, are given. Since these are the only values needed, $y = mx + b$, is called slope-intercept form. The slope and any point on the line must be known or produced to write an equation of a line in point-slope form. The form $y - y_1 = m(x - x_1)$ requires that only the slope, m, and a point on the line, (x_1, y_1), be given. These are the only values needed hence, the term point-slope form.

Write the equation of a line with slope $\dfrac{1}{2}$ and y-intercept 3.

For this line $m = \dfrac{1}{2}$ and $b = 3$. The equation $y = mx + b$ can be used since the values

of m and b are given. Replace m with $\dfrac{1}{2}$ and b with 3.

The result is $y = \dfrac{1}{2}x + 3$.

This is an equation for the line with slope $\dfrac{1}{2}$ and y-intercept 3.
The equation can be transformed to standard form, $Ax + By = C$, if desired.
Work to get the term with x on the left side of the equation to do so.
The process is on the following page.

Transform the equation of a line from slope-intercept form to standard form.

$$y = \frac{1}{2} x + 3$$ Subtract $\frac{1}{2} x$ from both side of the equation.

$$y - \frac{1}{2} x = \frac{1}{2} x + 3 - \frac{1}{2} x$$ Simplify. ($\frac{1}{2} x + 3 - \frac{1}{2} x = 3$)

$$y - \frac{1}{2} x = 3$$ Rearrange the terms on the left.

$$-\frac{1}{2} x + y = 3$$ Standard form.

The equation can be written without fractions by multiplying all terms by 2.

Multiplying $-\frac{1}{2}$ by 2 will result in a product of -1.

$$2 \left(-\frac{1}{2} \right) x + 2y = 3 \cdot 2$$

$-1x + 2y = 6$ Standard form with all integers.

Write the equation of a line that passes through the point (2, -3), with slope -2.

For this line $m = -2$ and (x_1, y_1) is (2, -3). The equation $y - y_1 = m(x - x_1)$ can be used since m, x_1 and y_1 are given.

Replace m with -2, x_1 with 2 and y_1 with -3.
$$y - (-3) = -2 (x - 2)$$
or
$$y + 3 = -2(x - 2)$$

The equation will be transformed to standard form on the next page.

Transform the equation of a line from point-slope form to standard form,

Standard is $Ax + By = C$.

$y + 3 = -2(x - 2)$	Distribute -2.
$y - 2 = -2x + 4$	Add $2x$ to both sides.
$y + 3 + 2x = -2x + 4 + 2x$	Simplify.
$2x + y + 3 = 4$	Subtract the 3 from both sides.
$2x + y + 3 - 3 = 4 - 3$	Simplify.
$2x + y = 1$	Standard form.

Write the equation of a line that contains the points (-2, -3) and (-5, 1).

The slope is not given in this case. The value of the slope is necessary and can be found when two points on a line are given. There is a formula used to calculate the slope of a line. For two points on a line (x_1, y_1) and (x_2, y_2), the slope, m, is found by

using the formula, $m = \dfrac{y_2 - y_1}{x_2 - x_1}$.

Match the points (-2, -3) and (-5, 1) with (x_1, y_1) and (x_2, y_2).

Next, replace x_1 with -2, y_1 with -3, x_2 with -5, and y_2 with -1.

$$m = \frac{1 - (-3)}{-5 - (-2)} \qquad \text{Simplify.}$$

$$m = \frac{4}{-3} = -\frac{4}{3}$$

The slope of the line is $-\dfrac{4}{3}$ and two points on the line are given.

The equation $y - y_1 = m(x - x_1)$ can be used since the slope and a point are given. In this case, two points are given. **Choose** the point to use for (x_1, y_1).

The slope of the line is $-\dfrac{4}{3}$ and the line passes through the points (-2, -3) and (-5, 1).

Since only one point is required to write an equation in point-slope form, **choose** one.

Use (-2, -3). The other point, (-5, 1) will work too. Make a **choice**.

In the equation $y - y_1 = m(x - x_1)$, replace m with $-\dfrac{4}{3}$, x_1 with -2 and y_1 with -3.

$$y - (-3) = -\dfrac{4}{3}\,[x - (-2)]$$

or

$$y + 3 = -\dfrac{4}{3}\,(x + 2)$$

The equation is in point-slope form.

Transform the equation of a line from point-slope form to standard form.

Standard is $Ax + By = C$.

$y + 3 = -\dfrac{4}{3}\,(x + 2)$	Distribute $-\dfrac{4}{3} \cdot 2 \cdot -\dfrac{4}{3} = -\dfrac{8}{3}$
$y + 3 = -\dfrac{4}{3}\,x - \dfrac{8}{3}$	Add $\dfrac{4}{3}\,x$ to both sides.
$y + 3 + \dfrac{4}{3}\,x = -\dfrac{4}{3}\,x - \dfrac{8}{3} + \dfrac{4}{3}\,x$	Simplify.
$\dfrac{4}{3}\,x + y + 3 = -\dfrac{8}{3}$	Subtract 3 from both sides.
$\dfrac{4}{3}\,x + y + 3 - 3 = -\dfrac{8}{3} - 3$	Simplify.
$\dfrac{4}{3}\,x + y = -\dfrac{17}{3}$	Standard form with fractions. Eliminate fractions by multiplying both sides (all terms) by 3.
$4x + 3y = -17$	Standard form with all integers.

If the point (-5, 1) had been used, the standard form would be the same.

Systems of Linear Equations in Two Variables

Two or more equations, with two or more unknowns (variables), form a system of equations. The two equations below form a system.

$$2x + 3y = 6$$
$$3x - 4y = -8$$

The solution to the system is the set of ordered pairs (x_1, y_1) that make **both** equations true. Is the ordered pair (3, 0) a solution for the system above? Replace x with 3 and y with 0. The ordered pair (3, 0) will make the first equation true since, $2(3) + 3(0) = 6$ is a true statement. The ordered pair (3, 0) will **not** make the second equation true since, $3(3) - 4(0) = -8$ is a false statement. The ordered pair (3, 0) is **not** a solution to the system because, (3, 0) does not make both equations true statements.

There are three methods that will be addressed for solving a system of linear equations in two variables. The first and least accurate is **graphing**. The advantage to graphing is that the solution is physically visible. The problem with graphing is that the graphs must be extremely precise and even then, there can be difficulties with reading the correct ordered pair(s). The other two are called **substitution** and **elimination**. These methods are accurate and involve solving linear equations in one variable. The processes of substitution and elimination both involve working the system down to a one variable equation. Solving a one variable equation was addressed in the text on pages 52-60.

The next page will show the solution to the system
$$2x + 3y = 6$$
$$3x - 4y = -8$$
found by graphing.

Solve the system by graphing.

$$2x + 3y = 6$$
$$3x - 4y = -8$$

The graph of the equation *2x + 3y = 6* shows the ordered pairs that will make the equation true. The graph of the equation *3x - 4y = -8* shows the ordered pairs that will make the equation true. Placing the graphs in the same coordinate plane will show the ordered pairs that satisfy both equations. Start by solving each equation for y. This will transform the equations to slope-intercept form.

$2x + 3y = 6$	$3x - 4y = -8$
$2x - 2x + 3y = 6 - 2x$	$3x - 3x - 4y = -8 - 3x$
$3y = -2x + 6$	$-4y = -3x - 8$
$\dfrac{3}{3}y = -\dfrac{2}{3}x + \dfrac{6}{3}$	$\dfrac{-4}{-4}y = \dfrac{-3}{-4}x - \dfrac{8}{-4}$
$y = -\dfrac{2}{3}x + 2$	$y = \dfrac{3}{4}x + 2$

Graph each equation in the same plane by using the slope and the y-intercept. Match each equation with the form $y = mx + b$, where *b* is the y-intercept and *m* is the slope.

The line with equation $y = -\dfrac{2}{3}x + 2$ has slope $-\dfrac{2}{3}$ and y-intercept 2.

The line with equation $y = -\dfrac{2}{3}x + 2$ has slope $\dfrac{3}{4}$ and y-intercept 2.

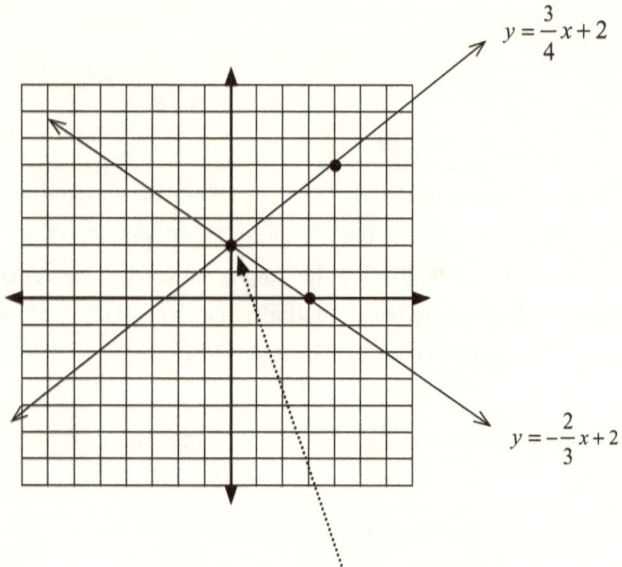

$y = \dfrac{3}{4}x + 2$

$y = -\dfrac{2}{3}x + 2$

Solution (0, 2) satisfies both equations

The solution to the system is the ordered pair associated with the point where the two lines intersect. To verify this, replace the x in each equation with 0, and the y with 2. The resulting equations should be true.

$$2x + 3y = 6 \qquad\qquad 3x - 4y = -8$$

$$2 \cdot 0 + 3 \cdot 2 = 6 \qquad\qquad 3 \cdot 0 - 4 \cdot 2 = -8$$

$$0 + 6 = 6 \qquad\qquad 0 - 8 = -8$$

Each statement is true.

The solution to the system, $\begin{array}{c} 2x + 3y = 6 \\ 3x - 4y = -8 \end{array}$, is the ordered pair (0, 2).

Solve the system by substitution.

$$2x + 3y = 6$$
$$3x - 4y = -8$$

Solving a system by substitution requires that one variable, in either equation, be isolated. The equation **chosen** and the variable **chosen** are left up to the individual. If the equation 2x+ 3y = 6 is solved for *x,* substitution will work. If the same equation is solved for *y,* the process will work. Similarly if either *x* or *y* is isolated in the equation 3x - 4y = -8, the process will work. Make a decision. Solve for *x* in the equation 2x + 3y = 6.

$$2x + 3y = 6$$

$$2x + 3y - 3y = 6 - 3y$$

$$2x = -3y + 6$$

$$\frac{2}{2}x = -\frac{3}{2}y + \frac{6}{2}$$

$$x = -\frac{3}{2}y + 3$$

The two equations to work with are now $x = -\frac{3}{2}y + 3$ and 3x - 4y = -8.

Next is the step where the substitution takes place.

Since *x* and $-\frac{3}{2}y + 3$ are the same, the *x* in 3x - 4y = -8 can be replaced

with $-\frac{3}{2}y + 3$.

Here is the result.

3x - 4y = -8

$$3\left(-\frac{3}{2}y + 3\right) - 4y = -8$$

replacement/substitution

Next, solve the equation $3\left(-\dfrac{3}{2}y+3\right)-4y=-8$, for y.

$3\left(-\dfrac{3}{2}y+3\right)-4y=-8$ Distribute

$-\dfrac{9}{2}y+9-4y=-8$ Collect like terms $\left(-\dfrac{9}{2}y+-4y=\ -\dfrac{9}{2}y+-\dfrac{8}{2}y=\ -\dfrac{17}{2}y\right)$

$-\dfrac{17}{2}y+9=-8$ Subtract 9 from both sides

$-\dfrac{17}{2}y+9-9=-8-9$ Simplify

$-\dfrac{17}{2}y=-17$ Divide both sides by $-\dfrac{17}{2}$

$\dfrac{-\dfrac{17}{2}y}{-\dfrac{17}{2}}=\dfrac{-17}{-\dfrac{17}{2}}$ Simplify $\left(\dfrac{-17}{-\dfrac{17}{2}}=-17\cdot-\dfrac{2}{17}=2\right)$

$y=2$ The value of y is 2.

y can be replaced in either equation with 2. This substitution will lead to the value of x. Replace y in $2x+3y=6$ with 2. It doesn't matter which equation is used. **Choose** one!

$2x+3\cdot2=6$ Solve for x.

$2x+6=6$

$2x+6-6=6-6$

$2x=0$

$x=0$ The value of x is 0.

The solution to the system, $\begin{aligned}2x+3y=6\\3x-4y=-8\end{aligned}$, is the ordered pair (0, 2).

Solve the system by elimination. Read this page twice!

$$2x + 3y = 6$$
$$3x - 4y = -8$$

The term elimination is used for a reason. The goal in this process is to eliminate one of the variables, to create an equation with only one variable. Solving a one variable equation was addressed in the text on pages 52-60.

First, create a situation where the coefficients on either x or y are opposites. **Choose** the variable to eliminate and follow the process described next. Eliminate x. The coefficients on x are 2 and 3. Find the Least Common Multiple (LCM) of 2 and 3. The LCM, of 2 and 3 is the smallest number that can be divided by 2 or 3. The LCM of 2 and 3 is 6. The coefficients of x will be changed to 6 and -6. Multiply both sides the equation 2x + 3y = 6 by 3. WHY 3? Make sure to multiply each term.

$$2x + 3y = 6$$

$$3(2x + 3y) = 3 \cdot 6$$

$$6x + 9y = 18$$

The coefficient of x is now 6. The coefficient of x in the second equation, 3x - 4y = -8 needs to be changed to -6. Multiply both sides of 3x - 4y = -8, by -2. WHY -2?

$$3x - 4y = -8$$

$$-2(3x - 4y) = (-8)(-2)$$

$$-6x + 8y = 16$$

The system, with different coefficients, is shown below.

$$6x + 9y = 18$$
$$-6x + 8y = 16$$

The next step is where the elimination occurs.

$$6x + 9y = 18$$
$$-6x + 8y = 16$$

Add the left members of the equation together, and do the same with the right sides.

$$6x + 9y = 18$$
$$\underline{-6x + 8y = 16}$$
$$0 + 17y = 34$$
$$17y = 34 \quad \text{Notice that } x \text{ has been eliminated.}$$

Solve the equation for y.

$$\frac{17}{17}y = \frac{34}{17}$$

$$y = 2$$

y can be replaced in either equation with 2. This substitution will lead to the value of x. It doesn't matter which equation is used. **Choose** one! Replace y in 2x +3y = 6 with 2.

$$2x + 3 \cdot 2 = 6 \quad \text{Solve for } x.$$
$$2x + 6 = 6$$
$$2x + 6 - 6 = 6 - 6$$
$$2x = 0$$
$$x = 0 \qquad \text{The value of x is 0.}$$

The solution to the system, $\begin{array}{l} 2x + 3y = 6 \\ 3x - 4y = -8 \end{array}$, is the ordered pair (0, 2).

The same system was solved in three different ways.
Each time the solution was and should have been the same.

**On the next page the system will be solved one more time by elimination.
This time y will be eliminated instead of x.**

Solve the system by elimination.

$$2x + 3y = 6$$
$$3x - 4y = -8$$

Eliminate y. The coefficients of y are 3 and -4. The LCM of 3 and -4 is 12. The coefficients of y need to be changed to 12 and -12.
Multiply both sides of the equation 2x + 3y = 6 by 4. The result is 8x + 12y = 24.
Multiply both sides of the equation 3x - 4y = -8 by 3. The result is 9x - 12y = -24.
The system written with different coefficients is below.

8x + 12y = 24
9x - 12y = -24

Add the left members and add the right members and solve for x.

$$
\begin{array}{r}
8x + 12y = 24 \\
9x - 12y = -24 \\
\hline
17x + 0 = 0 \\
17x = 0 \\
x = 0
\end{array}
$$

x can be replaced with 0 in either of the original equations to solve for y. Choose one.

$$2x + 3y = 6$$
$$2 \cdot 0 + 3y = 6$$
$$3y = 6$$
$$y = 2$$

Once again, the solution to the system, $\begin{array}{l} 2x + 3y = 6 \\ 3x - 4y = -8 \end{array}$, is the ordered pair (0, 2).

First, solve the system by graphing.
Then, solve the system by substitution.
Last, solve the system by elimination.
Try this before reading on.

$$3a + 2b = 10$$
$$-5a + 4b = -13$$

Solve the system by graphing.

$$3a + 2b = 10$$
$$-5a + 4b = -13$$

The solution(s) to the system is/are ordered pairs in the form (a, b). Unless instructed otherwise, use alphabetical order to choose which variable is used as the first coordinate. First, solve for b in both equations. b is treated like y from the previous examples.

$$3a + 2b - 3a = 10 - 3a \qquad\qquad -5a + 4b + 5a = -13 + 5a$$

$$2b = -3a + 10 \qquad\qquad\qquad 4b = 5a - 13$$

$$\frac{2}{2}b = \frac{-3}{2}a + \frac{10}{2} \qquad\qquad\qquad \frac{4}{4}b = \frac{5}{4}a - \frac{13}{4}$$

$$b = \frac{-3}{2}a + 5 \qquad\qquad\qquad b = \frac{5}{4}a - \frac{13}{4}$$

Next, graph the two equations in the same coordinate plane.

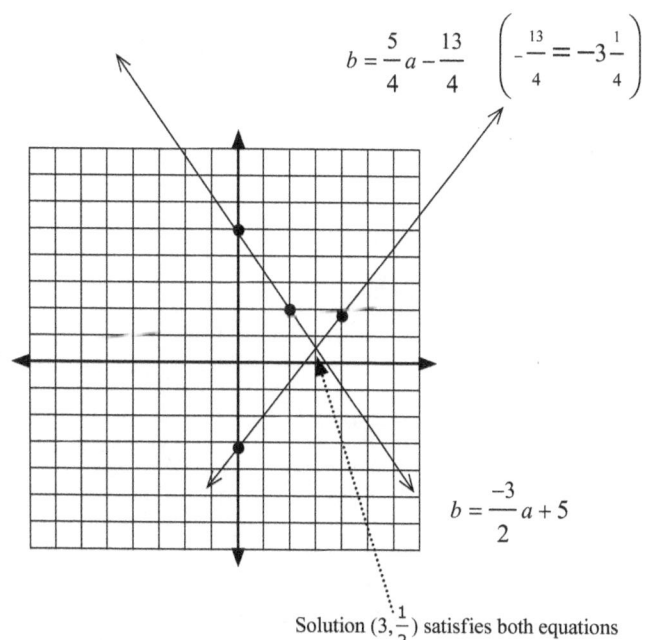

$$b = \frac{5}{4}a - \frac{13}{4} \qquad \left(-\frac{13}{4} = -3\frac{1}{4} \right)$$

$$b = \frac{-3}{2}a + 5$$

Solution $(3, \frac{1}{2})$ satisfies both equations

Solve the system by substitution.

$$3a + 2b = 10$$
$$-5a + 4b = -13$$

Solving a system by substitution requires that one variable, in either equation, be isolated. The equation **chosen** and the variable **chosen** are left up to the individual.

Solve for b in the first equation.

$$3a + 2b = 10$$
$$b = \frac{-3}{2}a + 5$$

Replace b in the second equation with $\frac{-3}{2}a + 5$.

$$b = \frac{-3}{2}a + 5$$

$$-5a + 4b = -13$$

$$-5a + 4\left(\frac{-3}{2}a + 5\right) = -13$$

Solve the equation for a.

$$-5a + 4\left(\frac{-3}{2}a + 5\right) = -13 \qquad \text{Distribute}$$
$$-5a + -6a + 20 = -13$$

$$-11a + 20 = -13$$
$$-11a + 20 - 20 = -13 - 20$$

$$-11a = -33$$
$$\frac{-11}{-11}a = \frac{-33}{-11}$$

$$a = 3$$

Next, solve for b by replacing a, in one of the original equations, with 3.

$$3a + 2b = 10$$

Choose one.   or

$$-5a + 4b = -13$$

Choose the equation to use for finding b.

$$3a + 2b = 10$$

$$3 \cdot 3 + 2b = 10$$

$$9 + 2b = 10$$

$$9 + 2b - 9 = 10 - 9$$

$$2b = 1$$

$$\frac{2}{2}b = \frac{1}{2}$$

$$b = \frac{1}{2}$$

The solution is $(3, \frac{1}{2})$.

Solve the system by elimination.

$$3a + 2b = 10$$
$$-5a + 4b = -13$$

Choose the variable to eliminate.

Eliminate b. Eliminating b will require less effort than eliminating a. Why? The coefficients of b are 2 and 4. The LCM is 4. Since the second equation already has a coefficient of 4 on b, simply multiply the first equation by -2 so that the coefficient on b will be -4.

$$-2(3a + 2b) = -2 \cdot 10$$
$$-6a - 4b = -20$$

The system, with different coefficients, is below.

$$-6a - 4b = -20$$
$$-5a + 4b = -13$$

Add the equations to eliminate b.

$$-6a - 4b = -20$$
$$\underline{-5a + 4b = -13}$$
$$-11a + 0 = -33$$
$$-11a = -33$$

and

$$a = 3$$

Next, solve for b by replacing a, in one of the original equations, with 3.

$$3a + 2b = 10$$

$$3 \cdot 3 + 2b = 10$$

$$9 + 2b = 10$$

$$9 + 2b - 9 = 10 - 9$$

$$2b = 1$$

$$\frac{2}{2}b = \frac{1}{2}$$

$$b = \frac{1}{2}$$

The solution is $(3, \frac{1}{2})$.

Systems can have no solutions or infinite (unlimited) solutions.

Systems can have no solutions or infinite (unlimited) solutions.

What will occur when solving a system that has **no solution**?

Graphing

Solving a system of two linear equations in two variables will produce a graph with two parallel lines. The lines will never intersect and there is no ordered pair that will satisfy both equations.

Substitution or Elimination

Solving a system of two linear equations in two variables will produce an equation with no variables. The equation will be false as well. When an equation such as $2 = 7$, or $-23 = 25$ is produced, there is no solution for the system.

What will occur when solving a system that has **infinite solutions**?

Graphing

Solving a system of two linear equations in two variables will produce a graph with one line. Each equation will produce the same graph. The line will show that all the ordered pairs that satisfy one equation will satisfy both equations.

Substitution or Elimination

Solving a system of two linear equations in two variables will produce an equation with no variables. The equation will be true. When an equation such as $2 = 2$, or $-23 = -23$ is produced, there are infinite solutions for the system. All the ordered pairs that satisfy one equation will satisfy both equations.

117

Solving Quadratic Equations Containing One Variable

An equation is a statement that contains an equal sign stating that two expressions represent or equal the same number. Examples of the equations are below.

$$2 + 2 = 4$$

$$w^2 = -7 \qquad \text{quadratic}$$

$$3r = 17 \qquad \text{linear}$$

$$11 \cdot 15 = 165$$

$$2n^2 - 7 = 0 \qquad \text{quadratic}$$

$$-\frac{t}{5} - 11 = 21 \qquad \text{linear}$$

$$3a^2 + 5a - 6 = 21 \qquad \text{quadratic}$$

$$k^2 + 3k = 12 \qquad \text{quadratic}$$

Notice that some of the equations have variables in them. These are called open sentences and one of the numbers is represented by a variable. When there is only one variable that appears in the equation and the variable is raised to the second power, the equation is called a quadratic equation in one variable.

The goal in **solving** an equation is to find the number(s) that the variable can be replaced with that will make the equation a true statement. A tool used to solve quadratic equations is called the quadratic formula. Using a formula often requires the replacement of variables with the value of the variables.

After making the replacement, if one variable is left, solving for its' value can be done.

Quadratic equations can be written in the form $ax^2 + bx + c = 0$, where a, b and c represent real numbers whose values are known. In other words, a, b and c represent constants. The a cannot represent zero. If a is zero then the equation is not quadratic. In the equation $3x^2 + 2x + 3 = 0$, $a = 3$, $b = 2$ and $c = 3$. Notice the 0 to the right the equal sign. **To identify a, b and c, the right side of the equal sign must be 0.**

In the equation $-4x^2 + \dfrac{1}{2}x - 4 = 0$, $a = -4$, $b = \dfrac{1}{2}$ and $c = -4$.

Solve a quadratic equation in the form $ax^2 + bx + c = 0$ for x, by using the **Quadratic Formula** shown below.

$$X = \frac{-b \pm \sqrt{b^2 - 4ac}}{2a}$$

The formula actually represents two solutions. Notice the $\pm$ symbol.

The $\pm$ is used to represent addition and subtraction. It indicates addition for one solution and subtraction for the second solution. The formula can be written in two parts as below.

$$X = \frac{-b + \sqrt{b^2 - 4ac}}{2a} \qquad\qquad X = \frac{-b - \sqrt{b^2 - 4ac}}{2a}$$

Use the Quadratic Formula to solve the equation below.

$$2x^2 + 4x - 3 = 0$$

Solve $2x^2 + 4x - 3 = 0$

First identify a, b and c. Use the general equation $ax^2 + bx + c = 0$.

$$a = 2, b = 4 \text{ and } c = -3$$

Next replace a, b and c in the Quadratic Formula with the values found.

$$X = \frac{-b \pm \sqrt{b^2 - 4ac}}{2a}$$

$$X = \frac{-(4) \pm \sqrt{(4)^2 - 4(2)(-3)}}{2(2)}$$

Use the order of operations to simplify the expression.

The **Order of Operations** is below.

First do work within grouping symbols.

Next do powers.

Next do multiplication and division from left to right.

Then, do addition and subtraction from left to right.

Simplify the expression under the radical ($\sqrt{}$) sign first.

$$\sqrt{(4)^2 - 4(2)(-3)} =$$

$$\sqrt{16 - 8(-3)} =$$

$$\sqrt{16 - (-24)} =$$

$$\sqrt{40}$$

Write the expression as shown on the next page.

$$X = \frac{-(4) \pm \sqrt{40}}{2(2)}$$

Simplify the denominator.

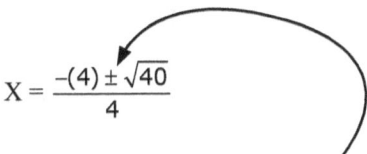

$$X = \frac{-(4) \pm \sqrt{40}}{4}$$

The expression above represents two values for x.

$$X = \frac{-(4) + \sqrt{40}}{4} \quad \text{and} \quad X = \frac{-(4) - \sqrt{40}}{4}$$

The difference in the values is produced by the addition in one, versus the subtraction in the other. If the expressions can be placed in lower terms, do so. The number 40 is not a perfect square and the $\sqrt{40}$ cannot be written as rational number. To get a better feel for the values of the two expressions, approximate the value of the $\sqrt{40}$ and simplify. Rounded to the nearest hundredth $\sqrt{40}$ is approximately 6.32. Replace $\sqrt{40}$ with 6.32.

$$\frac{-(4) + 6.32}{4} = 0.58 \qquad\qquad \frac{-(4) - 6.32}{4} = -2.58$$

Remember these are approximations. The exact values of x are represented in the expressions with radicals. So, x is approximately 0.58 or -2.58.

The symbol ~ means approximately.

$$x \sim 0.58 \text{ or } -2.58$$

The solution written without much commentary is on the next page.

Solve $2x^2 + 4x - 3 = 0$

$a = 2$, $b = 4$ and $c = -3$

Whenever following the process of problem solving, make sure you understand what has happened from one step to the next. For example, notice that $-\underline{24}$ in the third line below replaced the equivalent expression $4(2)(-3)$ in the second line.

$$X = \frac{-b \pm \sqrt{b^2 - 4ac}}{2a}$$

$$= \frac{-(4) \pm \sqrt{(4)^2 - 4(2)(-3)}}{2(2)}$$

$$= \frac{-(4) \pm \sqrt{16 - (-24)}}{4}$$

$$= \frac{-(4) \pm \sqrt{40}}{4}$$

$X \sim 0.58$ or -2.58

Use the Quadratic Formula to solve the equation below. $$\frac{1}{2}c^2 - 4c = 0$$	Identify a, b and c using the standard form $ax^2 + bx + c = 0$. The equation $\frac{1}{2}c^2 - 4c = 0$ can be written as $$\frac{1}{2}c^2 - 4c + 0 = 0 \cdot$$	The form showing three terms on the left side of the equal sign will provide an easier matchup with the standard form. The standard form $ax^2 + bx + c = 0$ also has three terms on the left.	The matchup of terms is as follows. ax^2 and $\frac{1}{2}c^2$ so, $a = \frac{1}{2}$ bx and $-4c$ so, $b = -4$ c and 0 so $c = 0$
$\mathbf{a} = \frac{1}{2}$, b= -4 and c= 0 Use the quadratic formula, $$\frac{-b \pm \sqrt{b^2 - 4ac}}{2a},$$ to solve for c.	Substitute the value of the variables into the formula. $C=$ $$\frac{-(-4) \pm \sqrt{(-4)^2 - 4\left(\frac{1}{2}\right)(0)}}{2\left(\frac{1}{2}\right)}$$	Simplify the expression. $$= \frac{4 \pm \sqrt{16 - 0}}{1}$$ $$= \frac{4 \pm \sqrt{16}}{1}$$	(Remember that $\pm$ means two answers are represented.) $$= \frac{4 \pm 4}{1}$$ $= 4+4$ or $4 - 4$ $C = 8$ or 0
Check by replacing c in the original equation with 8. Then replace it with 0.	$$\frac{1}{2}(8)^2 - 4(8) = 0$$ $$\frac{1}{2}(64) - 32 = 0$$	$32 - 32 = 0$ 8 works!	Check 0.
Use the Quadratic Formula to solve the equation below. $f^2 + 3f - 4 = 0$	Show work.		The answer is on the next page.

Solve $f^2 + 3f - 4 = 0$

$a = 1, b = 3$ an $c = -4$

$$\frac{-b \pm \sqrt{b^2 - 4ac}}{2a}$$

$$f = \frac{-3 \pm \sqrt{(3)^2 - 4(1)(-4)}}{2(1)}$$

$$= \frac{-3 \pm \sqrt{9 - (-16)}}{2}$$

$$= \frac{-3 \pm \sqrt{25}}{2}$$

$$= \frac{-3 \pm 5}{2}$$

$$= \frac{-3 + 5}{2} \text{ or } \frac{-3 - 5}{2}$$

$$= 1 \text{ or } -4$$

$$f = 1 \text{ or } -4$$

Check

$(1)^2 + 3(1) - 4 = 0$	$(-4)^2 + 3(-4) - 4 = 0$
$1 + 3 - 4 = 0$	$16 + (-12) - 4 = 0$
$4 - 4 = 0$	$4 - 4 = 0$
correct	correct

Use the Quadratic Formula to solve the equation below. $2e^2 = 3$ Zero must be the right member of the equation. Subtract 3 from both sides to accomplish this.	$2e^2 = 3$ $2e^2 - 3 = 3 - 3$ $2e^2 - 3 = 0$ Identify a, b and c using the standard form $ax^2 + bx + c = 0$. The equation $2e^2 - 3 = 0$ can be written as $2e^2 + 0e - 3 = 0$.	The form showing three terms on the left side of the equal sign will provide an easier matchup with the standard form. The standard form $ax^2 + bx + c = 0$ also has three terms on the left.	The matchup of terms is as follows. ax^2 and $2e^2$ so, a = 2 bx and 0e so, b = 0 c and -3 so c = -3
a=2, b=0 and c=-3 Use the Quadratic Formula, $$\frac{-b \pm \sqrt{b^2 - 4ac}}{2a},$$ to solve for e.	Substitute the value of the variables into the formula. $$C = \frac{-(0) \pm \sqrt{(0)^2 - 4(2)(-3)}}{2(2)}$$	Simplify the expression. $$= \frac{0 \pm \sqrt{0 - (-24)}}{4}$$ $$= \frac{0 \pm \sqrt{24}}{4}$$	(Remember that $\pm$ means two answers.) $$= \frac{\pm\sqrt{24}}{4}$$ $$= \frac{\sqrt{24}}{4} \text{ or } \frac{-\sqrt{24}}{4}$$ Simplify radicals. $$e = \frac{\sqrt{6}}{2} \text{ or } \frac{-\sqrt{6}}{2}$$ Use a calculator to approximate if needed.
Use the Quadratic Formula to solve the equation below. $f^2 = -6f - 2$ Remember zero must be the right member of the equation.	Hint: Add 6f to both sides and add 2 to both sides. Show work.		The answer is on the next page.

Solve $f^2 = -6f - 2$

$f^2 + 6f + 2 = -6f - 2 + 6f + 2$

$f^2 + 6f + 2 = 0$

$a = 1, \quad b = 6 \text{ and } c = 2$

$$f = \frac{-6 \pm \sqrt{(6)^2 - 4(1)(2)}}{2(1)}$$

$$= \frac{-6 \pm \sqrt{36 - 8}}{2}$$

$$= \frac{-6 \pm \sqrt{28}}{2}$$

$$= \frac{-6 \pm 2\sqrt{7}}{2}$$

$$= -3 \pm \sqrt{7}$$

$$= -3 + \sqrt{7} \text{ or } -3 - \sqrt{7}$$

An approximation to the nearest hundredth is below.

$$\sqrt{7} \sim 2.65$$

f ~ -3 + 2.65   or   -3 - 2.65

f ~ -0.35   or   -5.65

Translating Verbal to Math

Word problems present a challenge to most. **The important concept to remember is that in most cases, word problems relate to real life situations.** A skill that will help make solving a word problem easier is translating verbal phrases and sentences into their mathematical counterparts.

An Example: Translate the verbal sentence to a mathematical sentence.

The length of a rectangle is 3 more than the width.

Use variables to represent the length and the width. Normally w is used for width and l for length. Translating to math from a verbal phrase requires knowledge of the meaning of individual words. For example, the word is translates to =. More than translates to +. The statement with the verbal and mathematical symbols is as follows. The length/l of a rectangle is/= three more than/+ the width/w. The sentence written in mathematical symbols is below.

$$l = w + 3$$

In the mathematical sentence, unlike the verbal, the 3 follows w.

The reason for this is the original statement containing "more than w" implies the value three is added to w.

A situation where the translation above can be useful is as follows.

The Problem: The length of a rectangle is three more than the width. If the perimeter of the rectangle is 38cm, then what are the length and width?

Solution: The equation that relates the perimeter, the length and the width of a rectangle is $P = 2w + 2l$. P is the perimeter, w the width and l the length. Since the perimeter is 38, P can be replaced with 38. The equation with the substitution is $38 = 2w + 2l$. The problem with this equation is it contains two variables. The equation has many solutions for the length and width. Since there is a relation between l and w we can eliminate one of the variables. Substitute $w + 3$ for l and the result is an equation with one variable. The equation is $38 = 2w + 2(w + 3)$. This equation has one solution and the value of w can be found.

$$38 = 2w + 2(w + 3) \quad \text{Use the Distributive Property}$$

$$38 = 2w + 2w + 2 \cdot 3$$

$$38 = 4w + 6 \qquad \text{Solve for w.}$$

$$38 - 6 = 4w + 6 - 6$$

$$32 = 4w$$

$$\frac{32}{4} = \frac{4w}{4}$$

$$w = 8$$

The width of the rectangle is 8cm. To find the length use the equation $l = w + 3$. Since $w = 8$, substitute 8 for w.

$$l = 8 + 3$$
$$l = 11$$

The width of the rectangle is 8cm and the length is 11cm.

Word Problems

Word problems present a challenge to most. **The important concept to remember is that in most cases, word problems relate to real life situations.** Solving problems can often be made more manageable by relating the problem to a simpler situation. An example of this strategy is shown below. The solution is explained first and the explanation is followed by the format that should be used to respond to the directive.

The Problem: John has $10.00 to spend on prizes for his party. He has decided to purchase a combination of Snickers candy bars and Skittles. Each pack of Skittles cost $0.50 while each Snickers cost $0.60. If John buys 8 packs of Skittles, how many Snickers can he buy with the money left?

Write a mathematical equation to model the situation and respond to the question.

Solution: Determining the cost of a group of items is found by multiplying the price times the number of items. Finding the cost of 8 packs of Skittles is a simpler process than finding the solution to the problem above. Simply multiply the price of a pack of Skittles, $0.50, by the number of packs, 8. The result is $4.00. The amount to spend on Snickers is found by the same process.

The price of a Snickers, $0.60 is multiplied by the number of Snickers. Since the number of Snickers is not known, use a variable of your choice to represent the number. Use n and the cost of Snickers is represented by $0.60n.

KEVIN TUBBS

There are two groups of items being purchased. The total amount John will spend is the sum of the cost of both groups of items. Therefore, add the cost of Skittles, $0.50(8), and the cost of the Snickers, $0.60n, to find the total cost. The total cost is represented by the expression, $0.50(8) + $0.60n. Since John has $10.00, the total cost is equal to $10.00. The result is the equation below.

$$\$0.50(8) + \$0.60n = \$10.00$$

The units, $, do not need to appear in the equation.

$$0.50(8) + 0.60n = 10.00$$

Solve the equation for n and answer the question.

Here is the format that should be used to respond to the directions.

Let n = the number of Snickers that can be purchased.
(Always tell what the variable represents.)
The cost of Skittles plus/ + the cost of Snickers is/= ten dollars.
(Expressing thoughts in words can help create equations.)

$$0.50(8) + 0.60n = 10.00$$

$$4 + 0.6n = 10$$

$$4 + 0.6n - 4 = 10 - 4$$

$$0.6n = 6$$

$$\frac{0.6n}{0.6} = \frac{6}{0.6}$$

$$n = 10$$

John can buy 10 Snickers.

130

Solving problems can be more readily accomplished by using pictures, tables and/or graphs. Here is an example of the strategy.

The Problem: John left Fort Wayne for Purdue University at 2:00 p.m. John's father found John's phone on the couch one-half hour after John left. He immediately left to catch John to give John the phone. John has always been a responsible driver and travels at safe speeds. For this trip he averages 50mph. His father is not as responsible and will speed. He averages 75mph as he pursues John. At the rates the two travel, at what time will John's father catch John?

Write a mathematical equation to model the situation and respond to the question.

Solution: Pictures are often good for situations involving motion. This problem not only can be pictured, but it also can be modeled by a commonly used formula, $D = rt$. D represents the distance traveled, r represents the rate or speed traveled and t represents the time traveled. The formula says that the distance an object travels is equal to the rate the object travels at, multiplied by the length of time it travels.

A picture representing John's trip and his father's trip is shown on the next page.

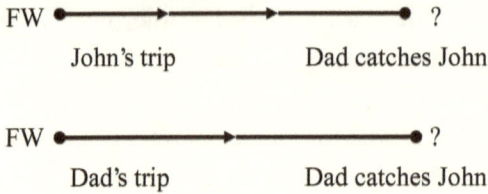

FW ●————→———→————● ?
John's trip   Dad catches John

FW ●————————→————————● ?
Dad's trip   Dad catches John

Notice that the trips begin and end at the same spot. John's father catches John so they both left the same location and they end in the same location. The fact that they left at different times does not affect the route they took or the distance they traveled.

The relationship between distance rate and time is the distance traveled equals the product of rate traveled and the time traveled. This relationship can be used to represent the trips that John and his father took, while comparing the two trips. Since John's rate was 50mph and his time was unknown, use 50 for rate and the variable t for time. Since Dad's rate was 75 and his time was unknown, use 75 for rate and (t -0.5) for time. Dad left one-half hour after John. If John travels t hours, then John's dad travels t -0.5 or one-half hour less than John. The illustration of the trips with the relationship of distance, rate and time is below.

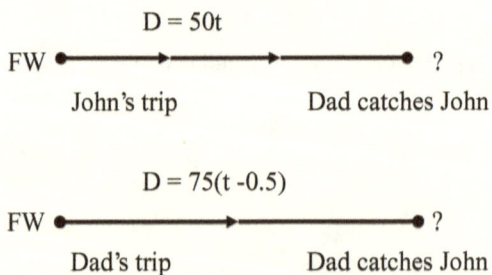

$D = 50t$

FW ●————→———→————● ?
John's trip   Dad catches John

$D = 75(t -0.5)$

FW ●————————→————————● ?
Dad's trip   Dad catches John

John and his dad traveled the same distance therefore the expressions for the product of rate and time represent the same numbers. In other words, 50t is the same as 75(t -0.5). Write an equation to state that the expressions/numbers are equal.

$$50t = 75(t - 0.5)$$

Solve the equation for t and the answer will be the length of time John traveled before his dad caught him.

$$50t = 75(t - 0.5)$$

$$50t = 75t - 75(0.5)$$

$$50t = 75t - 37.5$$

$$50t - 75t = 75t - 75t - 37.5$$

$$-25t = -37.5$$

$$\frac{-25t}{-25} = \frac{-37.5}{-25}$$

$$t = 1.5$$

John traveled 1.5 hours. Since John left at 2:00 pm, John's father catches him at 3:30 pm.

The problem might appear to be a long involved process. That is because of the commentary included. The format that should appear without comments is on the next page.

Let t be John's time driving

Let t -0.5 be dad's time driving

$$D = 50t$$

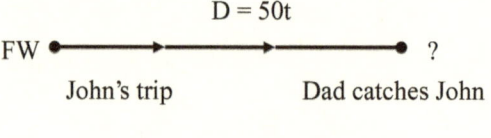

FW   John's trip   Dad catches John

$$D = 75(t -0.5)$$

FW   Dad's trip   Dad catches John ?

A table can be used <u>in place of or along with</u> of the picture.
Using tables helps to organize data.

	rate	time	distance
John's	50mph	t	50t
Dad's	75mph	t -0.5	75(t -0.5)

Using tables helps to organize data.

$$50t = D$$
$$75(t -0.5) = D$$

$$50t = 75(t - 0.5)$$

$$50t = 75t - 75(0.5)$$

$$50t = 75t - 37.5$$

$$50t - 75t = 75t - 75t - 37.5$$

$$-25t = -37.5$$

$$\frac{-25t}{-25} = \frac{-37.5}{-25}$$

$$t = 1.5$$

John drives for 1.5 hours before his dad catches him. His dad catches him at 3:30 pm.

Here is another example of problem solving with algebra.

The Problem: John has an honors banquet at Purdue tomorrow and realizes that his suit is still at home in Fort Wayne. John and his dad decide to drive toward each other, on their established route, so they can meet at a location for John to get his suit. They decide to pick a location that they can reach at the same time to avoid either one waiting for the other. Dad drives 60mph while John drives slower, 50mph. The distance from Fort Wayne to Purdue is 120 miles. If Dad leaves at 1:00 pm and John leaves at 1:15 pm how far from Fort Wayne will they meet?

Solution: John and Dad are driving toward each other to cover a set distance. An illustration of this situation is below.

120 miles total

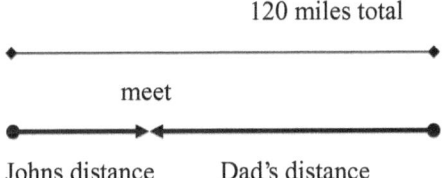

Johns distance   Dad's distance

In this situation John and his dad drive a total of 120 miles. The distances they drive individually are added together to total 120 miles. The rate John drives is 50mph and the time is unknown. Use t to represent the time John drives. His distance is then 50t. The rate Dad drives is 60mph and the time is unknown. Since Dad leaves 15 minutes or one-fourth hour before John, his time is $(t + 0.25)$. The distance Dad drives is $60(t + 0.25)$. The picture with the distance rate and times is on the next page.

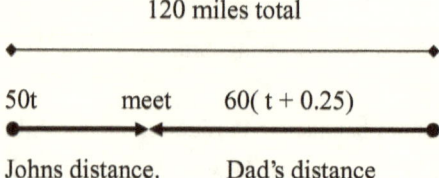

50t   meet   60(t + 0.25)

Johns distance.   Dad's distance

The sum of the distances John and his dad travel must equal 120. The equation stating this is below.

$$50t + 60(t + 0.25) = 120$$

Solve the equation for t, the time John travels.

$50t + 60(t + 0.25) = 120$

$50t + 60t + 60(0.25) = 120$

$110t + 15 = 120$

$110t + 15 - 15 = 120 - 15$

$110t = 105$

$\dfrac{110t}{110} = \dfrac{105}{110}$

$t \sim 0.95$ (Recall that $\sim$ means approxiamately)

0.95 is the number of hours John is traveling. The problem asks for the number of miles from Fort Wayne they are to meet. To find this distance, find the time Dad drives and multiply this value by the rate Dad drives. Since Dad drives 0.25 hours more than John, he drives 0.95 + 0.25 hours or 1.2 hours. The rate dad drives, 60mph multiplied by the time, 1.2 hours is about 72 miles. They will meet at a place 72 miles from Fort Wayne.

The format that should appear without comments is below.

Let t be the number of hours John travels.

Let t + 0.25 be the number of hours Dad travels.

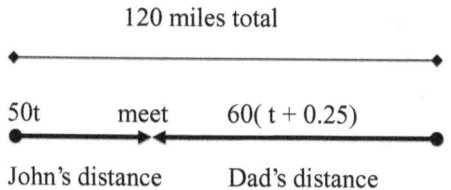

120 miles total

50t   meet   60(t + 0.25)

John's distance   Dad's distance

A table can be used <u>in place of or along with</u> the picture.
Using tables helps to organize data.

	rate	time	distance
John's	50mph	t	50t
Dad's	60mph	t + 0.25	60(t + 0.25)

Using tables helps to organize data.

$$50t + 60(t + 0.25) = 120$$

$$50t + 60t + 60(0.25) = 120$$

$$110t + 15 = 120$$

$$110t + 15 - 15 = 120 - 15$$

$$110t = 105$$

$$\frac{110t}{110} = \frac{105}{110}$$

$$t \sim 0.95 \quad \text{(Recall that} \sim \text{means approximately)}$$

Dad's time is t + 0.25 = 0.95 + 0.25 = 1.2 Distance from Fort Wayne = 60(1.20) = 72.

(Dad's rate multiplied by time)

They know they will meet 72 miles from Fort Wayne and can establish a meeting place.

The Problem: Two tables and four chairs cost a total of $360. Three tables and eight chairs cost a total of $620. What is the cost of one chair?

The Solution: The cost of a chair is unknown. The cost of a table is unknown. Variables are used to represent unknown values in mathematics.

Define the variables: Let t represent the cost of one table

Let c represent the cost of one chair

The total cost in the first situation, $360, is found by adding the cost of two tables, to the cost of four chairs. The cost of chairs is found by multiplying the number of chairs, 4, by the cost of one chair, c. The cost of tables is found by multiplying the number of tables, 2, by the cost of one table, t. $\underline{\ \ 2t + 4c = 360\ \ }$

The total cost in the second situation is found in the same manner, except different numbers are used. $\underline{3t + 8c = 620}$

Since there are two situations there are two equations. The two form a system of two linear equations in two variables.

Equations $\underline{\ \ 2t + 4c = 360 \quad 3t + 8c = 620}$

The system can be solved by graphing, substitution or elimination. The problem did not specify a method, so **choose one.** Use elimination to solve the system.

$2t + 4c = 360$
$3t + 8c = 620$

The problem requires c, the cost of a chair. Either variable can be eliminated, but it makes sense to eliminate t, the cost of a table since this value is not required.(see elimination pages 108-109)

2t + 4c = 360
3t + 8c = 620

To eliminate t, find the LCM of 2 and 3. The LCM is 6. The coefficients of t need to be change to 6 and -6. Multiply the first equation by 3 and the second by -3. Here is what should appear on paper.

$$3\left(2t + 4c\right) = 3 \cdot 360$$
$$-2\left(3t + 8c\right) = -2 \cdot 620$$

$$6t + 12c = 1080$$
$$\underline{-6t - 16c = -1240}$$
$$-4c = -160$$

$$\frac{-4}{-4}c = \frac{-160}{-4}$$

$$c = 40$$

Respond to the problem.

The cost of one chair is $40

The cost of a table can be found by substituting 40 for c in one of the original equations.

The Problem: A boat traveling with the current of the river moves at a speed of 65mph. The boat moves at a speed of 52mph against the same current. What is the rate of the current?

The speed the boat travels is affected by the current.

The boat's rate is not the 65mph or the 52 mph. The two values are the speed after the current has affected the boat's speed.

Let b represent the rate of the boat
Let c represent the rate of the current

The rate of the current will work with the rate of the boat when the boat travels with the current. (downstream) The current will help the boat travel faster.

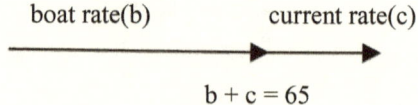

$$b + c = 65$$

The rate of the current will work against the rate of the boat when the boat travels against the current. (upstream) The current will slow the boat down.

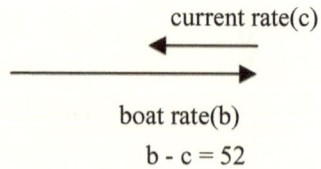

$$b - c = 52$$

Write a system of equations to model the problem.
$$b + c = 65 \qquad b - c = 52$$

Solve the system.

The system is ready for elimination without multiplying. Notice that the coefficients of c are opposites, 1 and -1. Simply add to eliminate c.

$$b + c = 65$$
$$\underline{b - c = 52}$$
$$2b = 117$$

$$\frac{2}{2}b = \frac{117}{2}$$

$$b = 58.5$$

The boat's rate is 58.5mph. The problem requires the rate of the current. Find the current's rate by substituting 58.5 for b in either equation.

$$b + c = 65$$

$$58.5 + c = 65$$

$$58.5 + c - 58.5 = 65 - 58.5$$

$$c = 6.5$$

The current's rate is 6.5mph.

The currents rate could have been found quicker by eliminating b instead of c. How could this be accomplished?

Quick Review of Adding Integers

Adding Integers with the same sign.	Adding Integers with different signs.

Adding integers with the same sign is like accumulating things. 5 + 6 means five ones plus six ones. This is a total of eleven ones. 5 + 6 = 11. On the other hand, -5 + -6 means five negative ones plus six negative ones. This is a total of eleven negative ones. -5 + -6 = -11
 	
When adding 5 + 6, two positives, think of the illustration below.
 	
   5   +   6   =   11
+ + + + + and + + + + + + = + + + + + + + + +
                         + + +
 	
When adding -5 + -6, two positives, think of the illustration below.
 	
  -5 + -6 =   -11
- - - - and - - - - - = - - - - - - - - - - -
	Adding integers with difference signs is like offsetting things. -5 + 6 means five negative ones plus 6 ones. Combining one and negative one is zero. By combining the two numbers five zeros will result, with one left over.
	
	When adding -5 + 6, different signs, think of the illustration below.
	
	   -5 + 6 = 1
	- - - - and + + + + + + = +
	
	For each negative paired with a positive the result is zero. There are enough negatives to pair with five positives.
	 -and + = 0, -and + = 0, -and + = 0, -and + = 0,
	 -and + = 0
	After pairing the negative ones with positive ones, one positive one is left over.
	
	                -5 + 6 = 1

Adding Integers with the same sign.	Adding Integers with different signs.
-15 + (-22) results in an accumulation of negatives. How many? 37 negatives. $$-15 + (-22) = -37$$ **The sum of two negative numbers, is negative.** -35 + -63 results in an accumulation of negatives. $$-35 + -63 = -98$$	-23 + 35 will result 23 zeros. The 23 negative ones will match with 23 positive ones. What will be left? 12 positive ones. $$-23 + 35 = 12$$ **The sign of the sum of two numbers with different signs is positive if there are more positives then negatives.** 23 + -35 will result 23 zeros. The 23 positive ones will match with 23 negative ones. What will be left? 12 negative ones. $$23 + -35 = -12$$ **The sign of the sum of two numbers with different signs is negative if there are more negatives then positives.**

Quick Review of Subtracting Integers

Subtracting Integers by Adding the Opposite.	Subtracting Integers by Adding the Opposite.
Subtraction means add the opposite. Subtracting two numbers is often made easier by converting the subtraction to addition. 6 - 11 means positive six minus positive eleven. To convert this expression to addition, add positive six to negative eleven. Instead of subtracting positive eleven, add negative eleven. <div align="center">6 - 11 =</div> <div align="center">6 + (-11) =</div><div align="center"><small>(six positives ones plus eleven negative ones)</small></div><div align="center">-5</div> <div align="center">6 - 11 = 6 + (-11) = -5</div>	-15 - 12 means negative fifteen minus positive twelve. Convert this to addition by adding negative twelve to the negative fifteen. Instead of subtracting twelve add negative twelve. <div align="center">-15 - 12 =</div> <div align="center">-15 + -12 =</div> <div align="center">-27</div> <div align="center">-15 - 12 = -15 + -12 = -27</div>
Subtracting Integers by Adding the Opposite.	Subtracting Integers by Adding the Opposite.
-21 - (-3) means negative twenty-one minus negative 3. Instead of subtracting negative three, add positive three. <div align="center">-21 - (-3) =</div> <div align="center">-21 + 3 =</div><div align="center"><small>(twenty-one negative ones plus three positive ones)</small></div><div align="center">-18</div> <div align="center">-21 - (-3) = - 21 + 3 = - 18</div>	33 - (-21) means thirty-three minus negative twenty-one. Instead of subtracting negative twenty-one, add positive twenty-one. <div align="center">33 - (-21) =</div> <div align="center">33 + 21 =</div> <div align="center">54</div> <div align="center">33 - (-21) = 33 + 21 = 54</div>

Quick Review of Fractions

Adding Fractions Common Denominator Needed	Subtracting Fractions Common Denominator Needed
$\dfrac{3}{8}+2\dfrac{1}{6}$ Write $2\dfrac{1}{6}$ as $\dfrac{13}{6}$. $\downarrow$ $\dfrac{3}{8}+\dfrac{13}{6}$ To find a common denominator list multiples of 8 (the larger) until $\downarrow$ you find one that is a multiple of 6. 8, 16, 24 Stop. Use 24 for the common denominator. $\dfrac{3\cdot3}{8\cdot3}+\dfrac{13\cdot4}{6\cdot4}$ $\downarrow$ $\dfrac{9}{24}+\dfrac{52}{24}$ $\downarrow$ $\dfrac{61}{24}$ or $2\dfrac{13}{24}$	$\dfrac{3}{14}-\dfrac{5}{6}$ List the multiples of 14 (the larger) until finding one that is a multiple $\downarrow$ of 6 to find a common denominator. 14, 28, 42 Stop. Use 42 for the common denominator. $\dfrac{3\cdot3}{14\cdot3}-\dfrac{5\cdot7}{6\cdot7}$ $\downarrow$ $\dfrac{9}{42}-\dfrac{35}{42}$ $\downarrow$ $\dfrac{9-35}{42}$ $\downarrow$ $\dfrac{-26}{42}=\dfrac{-13}{21}$ Place in lowest terms.
Multiplying Fractions Common Denominator **Not** Needed	Dividing Fractions Common Denominator **Not** Needed
$3\dfrac{1}{3}\cdot\dfrac{3}{8}$ Write mixed numbers as improper $\downarrow$ fractions. $\dfrac{10}{3}\cdot\dfrac{3}{8}$ Multiply numerators and multiply $\downarrow$ denominators. $\dfrac{10\cdot3}{3\cdot8}$ $\downarrow$ $\dfrac{30}{24}=\dfrac{5}{4}$ or $1\dfrac{1}{4}$ Write in lowest terms.	$\dfrac{4}{7}\div\dfrac{3}{11}$ Division means multiply by the reciprocal. $\downarrow$ Instead of dividing by $\dfrac{3}{11}$ multiply by $\dfrac{11}{3}$. $\dfrac{4}{7}\cdot\dfrac{11}{3}$ $\downarrow$ $\dfrac{4\cdot11}{7\cdot3}$ $\downarrow$ $\dfrac{44}{21}$ or $2\dfrac{2}{21}$

Quick Review of Powers

Negative exponents.	Negative exponents.
1. $5^{-3} = \dfrac{1}{5^3}$	3. $\left(\dfrac{2}{3}\right)^{-1} = \left(\dfrac{3}{2}\right)^{1} = \dfrac{3}{2}$
Negative exponents are used to represent reciprocals.	
5^{-3} means the reciprocal of 5^3.	4. $\dfrac{2d^{-4}c^2}{3^{-2}b^{-1}a^3} = \dfrac{3^2 \cdot 2b^1 c^2}{a^3 d^4} = \dfrac{18bc^2}{a^3 d^4}$
2. $\dfrac{1}{3^{-5}} = 3^5$	
The exponent zero.	**Multiplying Powers with the Same Base.**
1. $4^0 = 1$	1. $5^3 \cdot 5^4 = 5^{3+4} = 5^7$
Any number, except zero, to the zero power equals one.	When multiplying powers with the same base, add exponents.
2. $(33+5)^0 = 1$	$5^3 \cdot 5^4$ means $5 \cdot 5 \cdot 5 \cdot 5 \cdot 5 \cdot 5 \cdot 5$ which equals 5^7
3. 0^0 is not defined.	
4. $\left(6a^2 b^{-3} - 17\right)^0 = 1$	2. $3^2 x^4 \cdot 3^5 x^8 = 3^2 \cdot 3^5 x^4 x^8 =$ $3^{2+5} x^{4+8} = 3^7 x^{12}$

Continue on next page.

Quick Review of Powers (continued)

Dividing Powers with the Same Base.	Power of a Power(s).
1. $\dfrac{5^3}{5^4} = 5^{3-4} = 5^{-1} = \dfrac{1}{5}$ When dividing powers with the same base, subtract exponents. $\dfrac{5^3}{5^4}$ means $\dfrac{5\cdot5\cdot5}{5\cdot5\cdot5\cdot5}$ which equals $\dfrac{1}{5}$. 2. $\dfrac{3^2 x^8}{3^5 x^4} = 3^{2-5} x^{8-4} = 3^{-3} x^4 = \dfrac{x^4}{3^3}$	1. $(4^3)^2 = 4^{3\cdot2} = 4^6$ Multiply exponents when a power is raised to another power. $(4^3)^2$ means $4^3 \cdot 4^3$ which equals 4^{3+3} or 4^6. 2. $(2^6 x^3)^4 = 2^{6\cdot4} \cdot x^{3\cdot4} = 2^{24} x^{12}$ Powers distribute over multiplication and division.
Power of a Power(s).	**Assorted Powers. (Focus on like bases.)**
3. $\left(\dfrac{2r^3 y^2}{w^2 y^3}\right)^5 = \left(\dfrac{2r^3}{w^2 y}\right)^5 = \dfrac{2^5 r^{15}}{w^{10} y^5}$ $\left(\dfrac{2r^3 y^2}{w^2 y^3}\right)^5$ Simplify inside parenthesis. $\quad\quad \dfrac{y^2}{y^3} = \dfrac{1}{y}$ $\left(\dfrac{2r^3}{w^2 y}\right)^5$ Distribute exponent, 5, to $\quad\quad$ each factor in parenthesis. $\dfrac{2^5 r^{15}}{w^{10} y^5}$ Notice that 2 and y are raised to $\quad\quad$ the power of 5 since $2=2^1$ and $y=y^1$.	1. $\left(\dfrac{12 a^0 b^5 c^4}{19 1^0 b^{-2} c^3}\right) = \left(\dfrac{12 \cdot 1 b^5 b^2 c^{4-3}}{1}\right)$ $(a^0 = 1 \text{ and } 191^0 = 1)$ $= \left(\dfrac{12 b^7 c^1}{1}\right) = 12 b^7 c$ 2. $\left(13 x^2 y^{-3}\right)\left(13^{-1} x^3 y^2\right)\left(11 xyz\right)^0 =$ $\left(13 x^2 y^{-3}\right)\left(13^{-1} x^3 y^2\right)\cdot 1 =$ $13^{1+(-1)} x^{2+3} y^{-3+2} = 13^0 x^5 y^{-1} =$ $\dfrac{1 x^5}{y^1} = \dfrac{x^5}{y}$

Solving Proportions

Solve the Proportion	Solve the Proportion
$$\frac{2}{3} = \frac{5}{x}$$	$$\frac{4}{7} = \frac{5}{x+3}$$
Solve proportions by multiplying both sides of the equation, by each denominator. The process will eliminate the fraction. The quickest way to accomplish t he multiplication is using the cross product. The cross product is found by multiplying the upper left value, by the lower right. These are called the extremes. Set the product equal to the lower left value times the upper right value. These are called the means. Follow the procedure below.	$4(x+3)=7\times5$
	$4x+12=35$
	$4x+12-12=35-12$
	$4x=23$
$$\frac{extreme\ one}{mean\ one} = \frac{mean\ two}{extreme\ two}$$	$$\frac{4}{4}x = \frac{23}{4}$$
$extreme\ one \cdot extreme\ two = mean\ one \cdot mean\ two$	$$x = \frac{23}{4}$$
Solve the resulting equation.	**Solve the Proportion**
Solve the Proportion	$$\frac{4}{7} = \frac{x}{x+3}$$
$$\frac{2}{3} = \frac{5}{x}$$	$4(x+3) = 7x$
$2x = 5 \cdot 3$	$4x+12 = 7x$
$$\frac{2}{2}x = \frac{15}{2}$$	$4x+12-4x = 7x-4x$
	$12 = 3x$
$$x = \frac{15}{2}$$	$$\frac{12}{3} = \frac{3x}{3}$$
	$x = 4$